MW00895898

FROM TUSSOCKS TO TOURISTS

From Tussocks to Tourists

THE STORY OF THE CENTRAL CANTERBURY HIGH COUNTRY

David Relph

CANTERBURY UNIVERSITY PRESS

First published in 2007 by
CANTERBURY UNIVERSITY PRESS
University of Canterbury
Private Bag 4800, Christchurch
NEW ZEALAND
www.cup.canterbury.ac.nz

ISBN 978-1-877257-46-9

A catalogue record for this book is available from
the National Library of New Zealand

Printed in China through Bookbuilders

Front cover: Mountain biking in the Rakaia Valley.
Courtesy Dave Mitchell

Back cover: Mesopotamia in 1871.
Alexander Turnbull Library MNZ-0386-1/4F

Kea at Arthur's Pass; sweetbriar; tarn, tussocks and sheep, Waimakariri.

Half-title page: Mt Sunday in the Rakaia Valley with Mt D'Archiac in the distance.
Courtesy 4 X 4 New Zealand

Frontispiece: Lake Clearwater on a calm winter morning.

Title page: Lake Sarah, near Cass in the Waimakariri area.

CONTENTS

This book is dedicated to my father, Ted Relph, who as Chief Pastoral Lands Officer for the Lands Department in the South Island in the late 1940s and most of the 1950s was, I am told, highly regarded by high-country runholders for his fairness and dedication in balancing their interests with the need to maintain those of the Crown.

PREFACE AND ACKNOWLEDGEMENTS

'So what gives you the qualifications to write about the Canterbury high country?'

To be greeted with this question when meeting one of the elder statesmen among runholders was a great way to focus my mind. It was a fair question because although I grew up on a sheep station in the Rakaia Valley, my contacts with the region over the past 50 years had been minor and I was by now quite unknown to most people currently living there.

My answer to the question is that my childhood memories are filled with images such as wide-open landscapes covered with tussocks rippling in the wind; seemingly endless hot summer days during which I built stone dams in mountain streams; ice skating on Lake Ida; telegraph wires singing in fierce nor'west winds; hunting for paradise duck nests in the swamp; the joy of tramping through deep snow in gumboots and the resulting agony of thawing frozen feet in front of the stove.

A little later these memories include the smells and bustle of a busy shearing shed; of toiling up steep hillsides in the upper Waimakariri with a heavy pack; of one glorious summer climbing around Castle Hill Basin doing research for a university thesis; of driving a tractor on the terraces above the Lake Coleridge road in an attempt to sow turnips in a gale.

I don't know if these memories are sufficent qualifications to write about this region, but they certainly left me with a long-term love of the high country and an ambition to one day investigate it further. Over the next half-century my visits have been regular but brief. I hope that returning to research the area in detail after such a long period has given me a fresh perspective.

What started off as a journey down memory lane became a major voyage of discovery. A few aspects of high-country life have remained virtually unchanged over the past 50 years, but a great deal has altered dramatically and continues to do so, particularly as regards land management and the rise of tourism.

The location chosen as the focus for this book comprises the catchments of the three great Canterbury rivers – the Waimakariri, Rakaia and Rangitata – as well as the headwaters of the smaller Ashburton and Selwyn Rivers. It does not include the Hurunui and other rivers in North Canterbury, nor does it extend to the Waitaki catchment in the Mackenzie Basin. It was pointed out to me that 'Mid-Canterbury' excludes the Waimakariri area, so this book is about the high country of central Canterbury.

Why was this area chosen? First, covering the whole of the Canterbury high country seemed too daunting a project. The second reason is that my preliminary research suggested that the three major river valleys seemed to form a reasonably cohesive region, particularly in their natural environments. A key factor was the remarkably similar way that glaciation has shaped the huge valleys. Another unifying feature is the climate, particularly the way the valleys funnel the characteristic nor'west wind.

To the north, the rivers have not been carved by ice to the same extent, and although the Waitaki catchment to the south was modified by glaciation, this occurred in a much more open way.

Another consideration was my belief that Canterbury's high country has landscapes as impressive as those celebrated in Central Otago and the Mackenzie Country, and they deserved recognition.

This book discusses both the physical environment and the human impact on it, including early settlement, the era of the big sheep runs and the recent development of tourism.

The chapters on the natural environment – high-country landforms, climate and flora and fauna (native and exotic) – were relatively straightforward to write, partly because they are within my area of expertise and partly because these aspects have not changed greatly in the past 50 years. The main exception is the increasing entry of alien flora, both pests and pasture plants.

Researching the human presence the region was much more complex. While the history of the early exploration, settlement and development is relatively well documented, and much has been written about life on the traditional sheep runs, material on the later progress of the individual pastoral runs is often fragmentary and hard to find. In this respect Robert Logan's *Waimakariri: Canterbury's river of cold running water* was a treasure trove. Sadly, it has been long out of print and is difficult to obtain. Two more recently published books also provided invaluable material. John Chapman's *Stations of the Ashburton Gorge* (1999) is an excellent summary of the history of these properties. *West of Windwhistle*, published in 2005 by the Lake Coleridge Tourism Group, contains fascinating anecdotal material about life on the runs of the north side of the Rakaia.

It quickly became obvious to me that each of the 45 or so sheep runs in this region has a rich history and that there was no possibility of doing justice to the individual properties in a book of this scope. Though Chapter 7 is the longest in this book, only a few stations are discussed in any detail, and most have just a brief mention. They all deserve more, and I can only plead that constraints of both time and space precluded detailed treatment. I cannot speak too highly of the runholders of the relatively small number of properties I was been able to visit. Their unfailing assistance and willingness to share their knowledge was much appreciated.

The extent to which tourism has developed in the Canterbury high country in recent years was somewhat of a surprise to me, and turned out to be a relatively easy topic to research, thanks to the internet. Most of the commercial tourism operators have websites that provided a wealth of information. No doubt I have missed some, and to them I apologise. Those I met were enthusiastic about my project and willing to share their often extensive stock of excellent photos.

Researching and writing Chapter 10, on the issues of the administration and control of high-country land, was challenging because tenure issues can be contentious and the settlement process is ongoing. Much has been written about these matters, and I am also indebted to a number of runholders who offered generally very balanced (and sometimes strongly held) views. Mike Clare of the Canterbury Conservancy of the Department of Conservation provided me with clear assessments and assistance in identifying and providing maps of the areas under consideration. I have attempted to provide a balanced picture of the current situation by presenting the facts and the

This aerial photo looking west from the edge of the plains typifies the high country of central Canterbury, with its big, chunky mountains divided by the wide valleys of three great rivers. The Mt Hutt Range is on the left, with the Rakaia Valley to the right and Lake Coleridge in the far distance surrounded by mountains.
Courtesy Back Country New Zealand

Characteristic central Canterbury high country: in the distance, mountains cut by the wide valley of the Waimakariri River; in the middle ground, shingle terraces founded from glacial moraine; in the foreground, a tarn, tussocks, matagouri and sheep.

different viewpoints without making judgements.

The central Canterbury landscape is so striking that it seemed appropriate to conclude this book by discussing examples of the ways that it has influenced our culture, in particularly through art and, to some extent, writing. Sorting out and obtaining copies of the various works of art used in this book was an interesting, occasionally frustrating, but extremely rewarding task.

Producing *From Tussocks to Tourists* has been a real labour of love for me, and it will have succeeded if the book inspires others to explore this fascinating, if slightly neglected, part of New Zealand.

Many individuals and organisations contributed to this book, often unknowingly, and there are too many for all be acknowledged. I particularly wish to thank those who made significant contributions:

Mike Clare (Department of Conservation) for tenure review information. Tim Nolan for producing maps and diagrams. The University of Canterbury for information on its endowment lands.

The following runholders for hospitality and providing information and illustrations: James and Anna Guild (High Peak), Ben and Donna Todhunter (Cleardale), Tim and Anna Hutchinson (Double Hill), Mike and Karen Mears (Ryton), John Chapman (Inverary), Colin Drummond (Erewhon), Christine Fernyhough (Castle Hill) and Lyn Nell (Middle Rock).

The following tourism operators for information and photos: Warren and Marita Jowett (Bush & Beach Ecotours), particularly for natural history photos; Liam and Nigel (Back Country New Zealand); Les Cain (4 x 4 New Zealand); and Warner Raymond (Waimak Alpine Jet). Other photos were provided by John Bougen, Dave Mitchell (including the cover), Jack Creber and Nick Groves.

The following artists for their generous approval to reproduce their art: Diana Adams, Andrew Craig, Austen Deans, Michael Hight, Don McAra, Cristina Popovici and Ashley Smith.

For permission to use and assistance in obtaining historical paintings or photos: Corinne Crawley, Brian Hoult, Diana Lady Isaacs, Diane McKegg, B. Temple, Bruce Tinley, Alexander Turnbull Library, Ashburton District Council, Canterbury Museum, Christchurch Art Gallery, Hocken Library and the Waimakariri Art Collection Trust.

Jim Morris for permission to use his poetry, and John F. Cooke for his stories of Kb locomotives.

All photos other than mine are acknowledged in credits appearing with captions

Finally, thanks to my wife, Claire, for meticulous proofreading, for putting up with my prolonged mental absences and for enduring many hours of driving up long, dusty, high-country roads.

David Relph

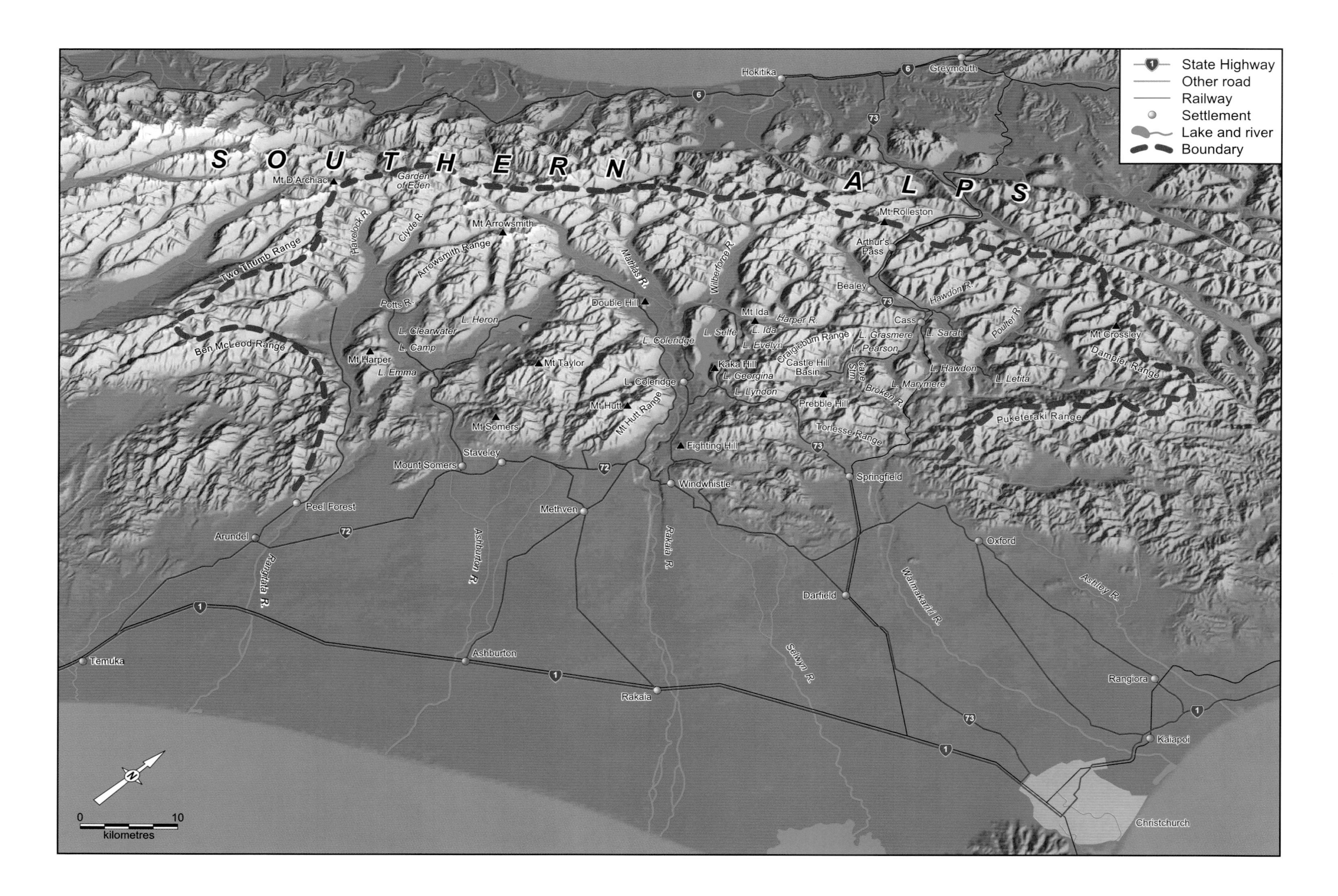
State Highway
Other road
Railway
Settlement
Lake and river
Boundary
SOUTHERN ALPS
Hokitika
Greymouth
Mt D'Archiac
Garden of Eden
Havelock R.
Clyde R.
Mt Arrowsmith
Arrowsmith Range
Two Thumb Range
Ben McLeod Range
Potts R.
L. Heron
L. Clearwater
L. Camp
Mt Harper
L. Emma
Mt Taylor
Mt Somers
Mathias R.
Double Hill
L. Coleridge
Mt Hutt
Mt Hutt Range
Wilberforce R.
L. Selfe
Mt Ida
L. Ida
Harper R.
L. Evelyn
Kaka Hill
L. Georgina
L. Lyndon
Craigieburn Range
Castle Hill Basin
Cave Stm
Prebble Hill
Fighting Hill
Torlesse Range
Broken R.
L. Grasmere
L. Pearson
L. Marymere
L. Hawdon
L. Sarah
Cass
Bealey
Arthur's Pass
Mt Rolleston
Hawdon R.
Poulter R.
L. Letitia
Mt Crossley
Dampier Range
Puketeraki Range
Staveley
Mount Somers
Windwhistle
Springfield
Peel Forest
Methven
Arundel
Oxford
Rangitata R.
Ashburton R.
Rakaia R.
Selwyn R.
Waimakariri R.
Ashley R.
Darfield
Temuka
Ashburton
Rakaia
Rangiora
Kaiapoi
Christchurch
0
10
kilometres

CHAPTER ONE

Ice and Shingle

Origins of the landforms

This chapter explains the origins of the central Canterbury high-country region and, using photographs, maps and other illustrations, describes how to interpret its typical landforms.

From any vantage point in the catchment areas of the three great Canterbury rivers, spectacular mountain peaks dominate the horizon in every direction. There are some broad, open basins and wide, braided riverbeds, but it is the surrounding mountains that impose their presence on all parts of the region. They stretch for 60 to 70 kilometres from the Southern Alps in the northwest to the Canterbury Plains in the southeast. These highlands are of rather similar appearance and height, 1,500–2,000 metres, and are a jumble of steep-sided mountains, their upper slopes covered in grey shingle (scree) that extends as ribbons down into the tussocks on the lower slopes.

The northwestern boundary of the region we are exploring is made up of the main range of the Southern Alps. This forms a continuous line of jagged mountains, about 2,000 metres high north of the Waimakariri and somewhat higher further south, culminating around Mt D'Archiac in a cluster of impressive peaks reaching over 2,600 metres.

To the east of the Main Divide is the 'front country', a series of ranges forming the boundary with the Canterbury Plains. While often steep and deeply dissected by streams, ranges such as the Malvern Hills, Mt Peel and Mt Somers have somewhat more rounded peaks and generally do not rise very much above 1,500 metres.

Behind the front ranges lie several broad basins, such as at Castle Hill and Lake Heron, but these systems are also surrounded and dominated by mountains.

Opposite: The central Canterbury high-country area explored in *From Tussocks to Tourists*.
Cartography by Tim Nolan

Below the ever-present peaks, the defining features of the area are the valleys of three great rivers, the Waimakariri, Rakaia and Rangitata. Each of these has carved a wide, almost straight bed, like a giant motorway cutting right across the jumble of mountains. The catchments of the three rivers, which stretch over 130 kilometres northeast to southwest and cover close to 8,000 square kilometres, have a common geological history and have been shaped by similar forces.

Greywacke everywhere

Pick up a piece of rock almost anywhere in this area – be it a round pebble from a riverbed or a sharp, angular fragment from high up a shingle slide – and the chances are that it will be a lump of greywacke.

This rock, omnipresent in over 90 per cent of the region, is a fine-grained sandstone or mudstone that has been subjected to great pressure and become very hard. Greywacke is rather nondescript in appearance – usually grey, though sometimes creamy or even reddish in colour and often cut by streaks of white quartz. The class of rocks known as Torlesse

Greywacke rocks, one jagged and the other rounded by water.

greywackes (after the Torlesse Range) varies from lighter, larger-grained sandstones to very fine-grained, dark mudstones, known as argillite, often in alternating layers.

The material making up these rocks was deposited in shallow seas between 250 and 100 million years ago. Appearing in outcrops as jumbled, twisted and broken layers, the rocks reflect a long history of uplift, erosion, compression and faulting. They now form the underlying rock of most of New Zealand.

Other rock types in the region are found in very small and isolated locations. Younger marine sediments were deposited above the greywacke, and on the floors of some valleys and basins they remain because, being in a hollow, they have not been stripped away by erosion. Most spectacular of these sediments are the distinctively shaped outcrops of limestone in the Castle Hill Basin.

There are a number of other smaller pockets of sedimentary strata exposed in the eastern foothills, including further limestone outcrops around Mt Somers, and several small, fractured coal seams.

A few areas of volcanic rocks are also exposed in the hill country, including the andesite and rhyolite that make up the western Malvern Hills and around High Peak. Further patches of rhyolite, containing small garnets, are exposed in the Rakaia Gorge near the bridge. Rhyolites and andesites also make up much of the Clent Hills, Mt Somers and Mt Alford around the Ashburton Gorge, and there are outcrops in the hills to the south of the Rangitata Gorge. These are the result of volcanic activity in the early Cretaceous period, up to 130 million years ago. Some of these rocks, particularly those from Mt Somers, are sought after by gemstone hunters and they may also be found polished by the sea on the beach at Birdlings Flat, Kaitorete Spit, having been carried down the Rakaia River.

Origins

In order to trace the origin of the Torlesse greywackes, we have to travel back over 300 million years to the Carboniferous period. At that time vast quantities of grey sand and black mud were being eroded away from the great southern land mass we now call Gondwana and deposited in the sea to the east in an area that would eventually become New Zealand. The layers of sediments were then involved in the roller-coaster ride of compression, uplift, faulting and erosion that went on for the next 250 million years, and which in fact continues today.

A period of mountain building, the Rangitata Orogeny, resulting from the movement of crustal blocks, converted the layers of mud and sand into hard greywacke and argillite rocks, which were pushed up to become the cores of the present mountains of the Canterbury high country. By 100 million years ago this era of mountain building had ceased, and about that time the volcanic activity occurred that left its traces in the eastern foothills.

As the mountains slowly rose, the forces of erosion immediately began to act on them. For a vast period the highlands were slowly worn away as the area that would become New Zealand drifted away from Australia. By 35 million years ago this was again largely under the sea and much of the greywacke rock was being covered by further marine sediments, which remain in isolated pockets such as the limestones of Castle Hill.

The Maori origin story

The story of the creation of the Southern Alps, as related by Ngai Tahu, the southern Maori people, begins before there was any sign of land.

Aoraki and three of his brothers were sons of Raki, the sky father. They came down from the sky and explored the sea in their waka (canoe), looking unsuccessfully for land. Unable to return to their celestial home because of the failure of a karakia (incantation), they ran aground on a submerged reef.

As the waka settled into the water it turned to stone and earth, forming the South Island (Te Waka o Aoraki). The waka listed to the east and the higher side became the Southern Alps (Ka Tiritiri a te Moana) with the four brothers being turned into stone and forming peaks as they clambered away from the sea.

Aoraki became Aotearoa's highest mountain. Although later named Mt Cook after the great English mariner, recently it has been appropriately renamed Aoraki/Mt Cook.

The later shaping of the land was largely the result of the deeds of Tu Te Raki Whanoa, who wielded an adze to carve out lakes and fiords.

During Miocene times, from 25 to 5 million years ago, further changes that foreshadowed the present New Zealand landforms were beginning. The Earth's crust in the southwest Pacific was moving, and the land that was to be New Zealand was straddling the boundary between two tectonic plates. As a result, the area became increasingly unstable and another period of uplift (the Kaikoura Orogeny) began.

Not only were the mountains being pushed up, but there was massive northeast–southwest sideways movement along the boundary of the Pacific and Australian Plates. The stresses created there caused buckling and shearing of the rocks and pushed up the Southern Alps, which run along the eastern side of the Alpine Fault.

The extent of this uplift was extraordinary – up to eight kilometres along the western margin. Extending east, the buried greywacke rocks were pushed up to form mountains again. Quite extensive faulting took place in that region, particularly parallel to the main Alpine Fault, which resulted in some blocks being forced up and others lowered. A good example of this is the uplifted Torlesse Range alongside the down-faulted Castle Hill Basin.

Cleft, Andrew Craig's painting based on a locality in the Rakaia Valley, captures the feeling of this timeless land, with mountains being forced up by the movement of tectonic plates and then inexorably worn down by the forces of erosion.

Private collection, courtesy Andrew Craig

After being subjected to enormous periods of earth movement, it is hardly surprising that greywacke rocks often lie in tortured and twisted beds.

As this enormous uplift was taking place the forces of erosion immediately began working on the mountains. The higher parts in particular were eroded away, exposing the greywacke again. Remnants of overlying marine sediments occur only in the downfaulted basins such as at Castle Hill and in a few lines of faulting such as the Esk and Harper Valleys.

Along with marine sediments, there are also a number of very fractured beds of coal, formed from ancient forests, at Mt Somers, the Acheron Stream in the Rakaia Valley, and Broken River in the Waimakariri catchment.

By five million years ago the broad pattern of the landforms was quite similar to the present day, with jumbled ranges of mountains extending east from the Southern Alps to the sea, which at that time covered most of what is now the Canterbury Plains. The intermontane basins created by faulting were present and it is possible that the big rivers wound through the hills in much the same courses as today. But the details of the landscape were very different. The surface of the mountains and valleys was yet to be acted on to produce the characteristics of the current landscape.

The 'Ice Age'

Canterbury's high country owes its most distinctive features to the event known as the 'Ice Age'. About 2.4 million years ago the seas around New Zealand began to cool, the climate chilled and ice built up, particularly in the highlands of the South Island. From this time through to the present, there have been a number of climate oscillations, resulting in a series of ice advances, with milder spells in between.

It is not easy to correlate the age and sequence of the different periods of glaciation, but it is clear that in the past 500,000 years there have been at least four periods of ice advance, and we are in one of the interglacial periods right now. So the 'Ice Age' was not one ice advance but a whole series. This is important when looking at the effects of the ice, because the erosion that took place in the interglacial periods was just as influential as the actual ice advances in determining the present form of the land. Colder climates persisted until about 10,000 years ago, when conditions began to warm up again.

During the period of the Waimaunga Glaciation (280,000–220,000 years ago) most of the central Canterbury high country appears to have been covered in an almost continuous ice sheet. The two more recent glacial periods, the Waimea (180,000–125,000 years ago) and Otira (75,000–14,500 years ago), seem to have been somewhat less extensive and have left well-preserved moraines further back in the mountains.

At its greatest extent the ice sheet must have been an impressive sight. It was centred along the Southern Alps, where it covered all but the tops of the higher peaks, and it spread eastward to cover perhaps 75

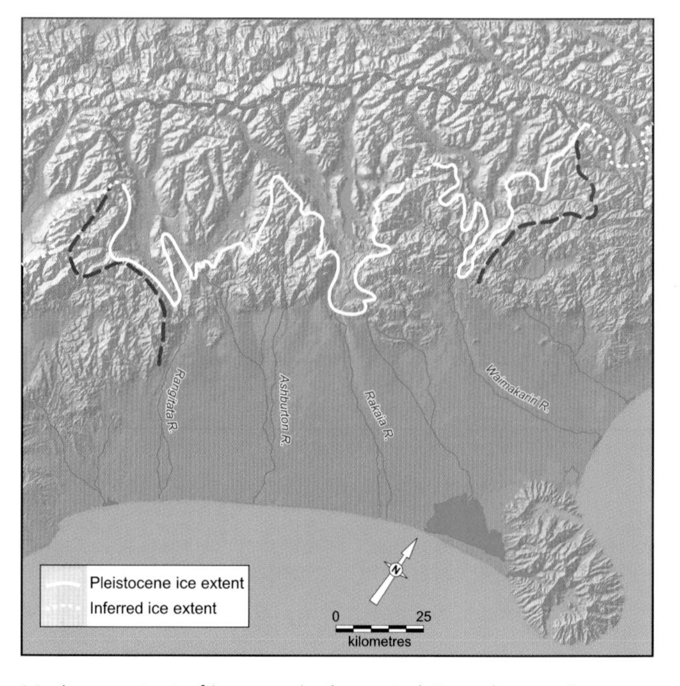

Maximum extent of ice cover in the central Canterbury region.
Cartography by Tim Nolan

From the Canterbury Plains range upon range of mountains extend back to the Southern Alps, the 'spine' of the South Island.
Courtesy Back Country New Zealand

per cent of the high country in massive tongues of ice hundreds of metres thick.

The Rangitata and Waimakariri Glaciers were halted by the eastern hill country, but at its greatest advance, about 20,000 years ago, the huge Rakaia Glacier extended right through the front ranges and onto the Canterbury Plains, where its snout protruded as much as seven to eight kilometres. Since then the ice has retreated and advanced at least three times, on each occasion extending a shorter distance and leaving huge amounts of moraine behind.

The effects of the ice on the underlying landforms was immense. The most recent glaciation, the Otira, has left the greatest mark on the surface of the land. It has done this in two ways: the mountains were sculpted by the erosive effects of the ice, and then, as the ice retreated, erosion occurred, depositing vast amounts of shingle in the valleys. The full extent of the previous ice covering is disguised by these later gravel deposits.

The Canterbury Plains

Although detailed study of the Canterbury Plains is beyond the scope of this book, it is worth remembering that they are the result of the outpouring and build-up of outwash gravels from the three great rivers. Vast fans of sedment built up and eventually coalesced to form a vast plain, over 60 kilometres across and stretching at least 150 kilometres from north to south. The 'joins' between the three fans is approximately the paths taken by the Selwyn and Ashburton Rivers, which drain the front ranges.

Looking at the landforms

Although most of the present landscapes of the three great rivers owe much to the ice ages, they have been affected in different ways. They also cover a very large area, so any attempt to describe their features throughout the whole region would be a huge and tedious task. Instead, we take a closer look at a sample of sites that represent the main varieties.

The map on page 10 shows the topography present in this area. The descriptions that follow are of the range of sites that are visible from the five roads that penetrate the high country. As you drive up any of these roads it becomes apparent that there are four fairly clearly defined zones, each with its characteristic landforms.

This view of the upper Rakaia illustrates many of the typical features of the glacial landforms that have shaped the three great Canterbury river valleys. Ancient glaciers have trimmed, straightened and steepened the valley sides, and the mass of shingle filling the floor is the result of rapid erosion following glacial retreat. In the middle ground the surface of Prospect Hill has been carved by ice action, and a tarn occupies a hollow on the irregular surface.
Courtesy Back Country New Zealand

The front ranges
The lower ranges bordering on the Canterbury Plains were never covered by ice sheets, so this country is shaped by river erosion. Mt Peel and Mt Somers, the Big Ben Range and, particularly, the Torlesse Range feature large and rugged mountains that are quite steeply and irregularly dissected, generally forming valleys V-shaped in profile.

The range-and-basin country
The large open basins inland from the front ranges were generally formed by downfaulting and are surrounded by higher block mountains. The floor of the Castle Hill Basin, with its younger limestone rocks, was not glaciated, but the Cass and Lake Heron Basins both contain large amounts of moraine deposited by glaciers.

The three great glacially carved river valleys
The inland valleys of the Waimakariri, Rakaia and Rangitata are large and spectacular regions. They were widened and straightened by huge glaciers, and their floors partly filled with shingle. At the downstream end of the Rakaia Valley there are dumps of glacial moraine and terraces formed by water transporting and depositing the gravel. These features are readily seen from the road to Lake Coleridge, and at the Rakaia Gorge.

The alpine highlands
The high ranges along the edge of the Southern Alps are characterised by very steep-sided valleys, cirques, jagged arêtes, rocky surfaces and other landforms that have been extensively carved out by ice.

Landforms of the Waimakariri Valley

The Waimakariri is the best known of the three big river valleys, with road and rail links right across to the West Coast, in spite of the formidable barrier of Mt Torlesse making access difficult over Porters Pass. Both routes are among the most varied and spectacular in the South Island, passing through some obvious examples of the four distinctive landscape types.

Above: Approaching the Torlesse Range from Springfield.

Right: Valleys carved by water into the shingle on the Torlesse Range. The higher slopes are now part of the Korowai/Torlesse Tussocklands Park.

Torlesse and Big Ben Ranges

Block mountains deeply dissected by rivers

Most people approach the Canterbury high country from the east by the main West Coast road (SH73). I'm sure I am not alone in feeling a tingle of anticipation as the massive south face of the Torlesse Range, with its distinctive notch in the ridge, looms larger and larger. From the Kowai River flats it becomes clear that the slopes of the range are heavily dissected by streams that have carved steep gullies, beginning in the scree of the summit ridges and funnelling into narrow, shingle-filled trenches that cut through the forest or tussock of the lower slopes.

The Torlesse is the largest of the front ranges, and it not only effectively blocked the Waimakariri Glacier from reaching the Canterbury Plains but forced the river to cut a deep gorge through the range, which the railway follows today. This massive greywacke block has been gradually uplifted along the line of faults on its northwestern side. Seen from the road ascending Porters Pass, the hills on the left have the typical appearance of mountains eroded into irregular steep valleys by rivers.

Although the Big Ben Range at its peak reaches over 1,600 metres, these hills are generally lower than the Torlesse, with more rounded tops.

Castle Hill Basin (Kura Tawhiti)

Downfaulted basin with distinctive sedimentary limestone strata

SH73 descends from Porters Pass over gravel terraces and shingle fans to enter Castle Hill Basin, an almost surreal landscape and perhaps the most distinctive part of the whole high country. This large inter-montane basin comprises a downfaulted block surrounded by raised greywacke highlands. The southeast is bounded by the Torlesse Range. To the west the Craigieburn Range forms another massive block of mountains, while to the northeast are the smaller Broken Hills.

The floor of the basin is made up of Tertiary sediments that, because they are in a downfaulted hollow, have been preserved from being eroded away. The result is a landscape that deserves detailed explanation. The very distinctive outcrops of limestone may seem a confused jumble in various parts of the basin, but there is a pattern about them. From a vantage point in the basin (such as one of the spurs running down from the Craigieburn Range), this pattern can be picked out.

The layers, or strata, of sediments are like a series of dinner plates covering the basin floor. The bottom plate is a layer of sands and silts with some coal bands exposed on the eastern and northern edges.

Limestone formations at Castle Hill with the Craigieburn Range behind.

These sheep-like limestone remnants illustrate how Flock Hill got its name. Their spectacular shapes formed part of the set for battle scenes in the film *The Lion, the Witch and the Wardrobe*.

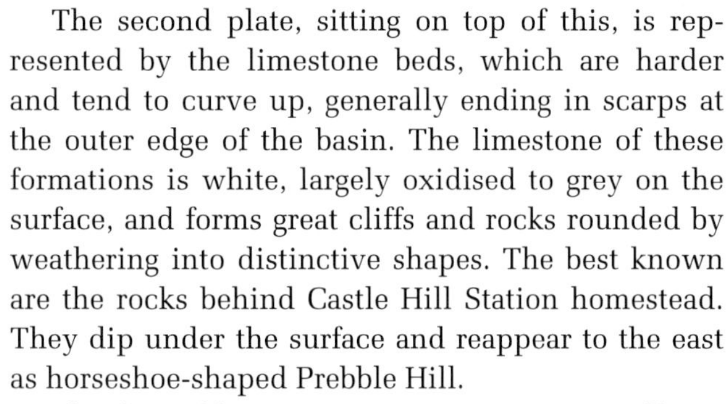

The second plate, sitting on top of this, is represented by the limestone beds, which are harder and tend to curve up, generally ending in scarps at the outer edge of the basin. The limestone of these formations is white, largely oxidised to grey on the surface, and forms great cliffs and rocks rounded by weathering into distinctive shapes. The best known are the rocks behind Castle Hill Station homestead. They dip under the surface and reappear to the east as horseshoe-shaped Prebble Hill.

This line of limestone reappears as Gorge Hill, cut by the Broken and Porter Rivers, then on the north side of Broken River as Flock Hill, which also has a steep scarp facing east. On the slope facing the road the large limestone rocks resemble a huge flock of sheep. A tunnel easily accessible from the road has been cut by Cave Stream to reach Broken River.

Complicating the pattern of landforms further, our model includes a third plate – another layer of gritty sandstones – sitting on the central part of the limestone layer. On top of this again are irregular

View looking northwest to the Craigieburn Range across Castle Hill Basin, showing gravel terraces cut by the Broken River.

deposits of river gravels that arrived much more recently. A series of streams draining the Craigieburn Range have cut through the upper layers of gravels, forming quite deep gullies with terraces on each side, across which the road winds up and down. These streams, including the Porter River, Whitewater Creek, Thomas River and Hogsback Creek, all drain into the Broken River, which flows east to the Waimakariri. Wherever the rivers cut through the limestone they have created spectacular gorges.

Right: Entrance to the nearly 600-metre-long tunnel created by Cave Stream through limestone.

Below: Diana Adams' painting *Boulders – Castle Hill* emphasises the surreal nature of this landscape.
Courtesy Diana Adams

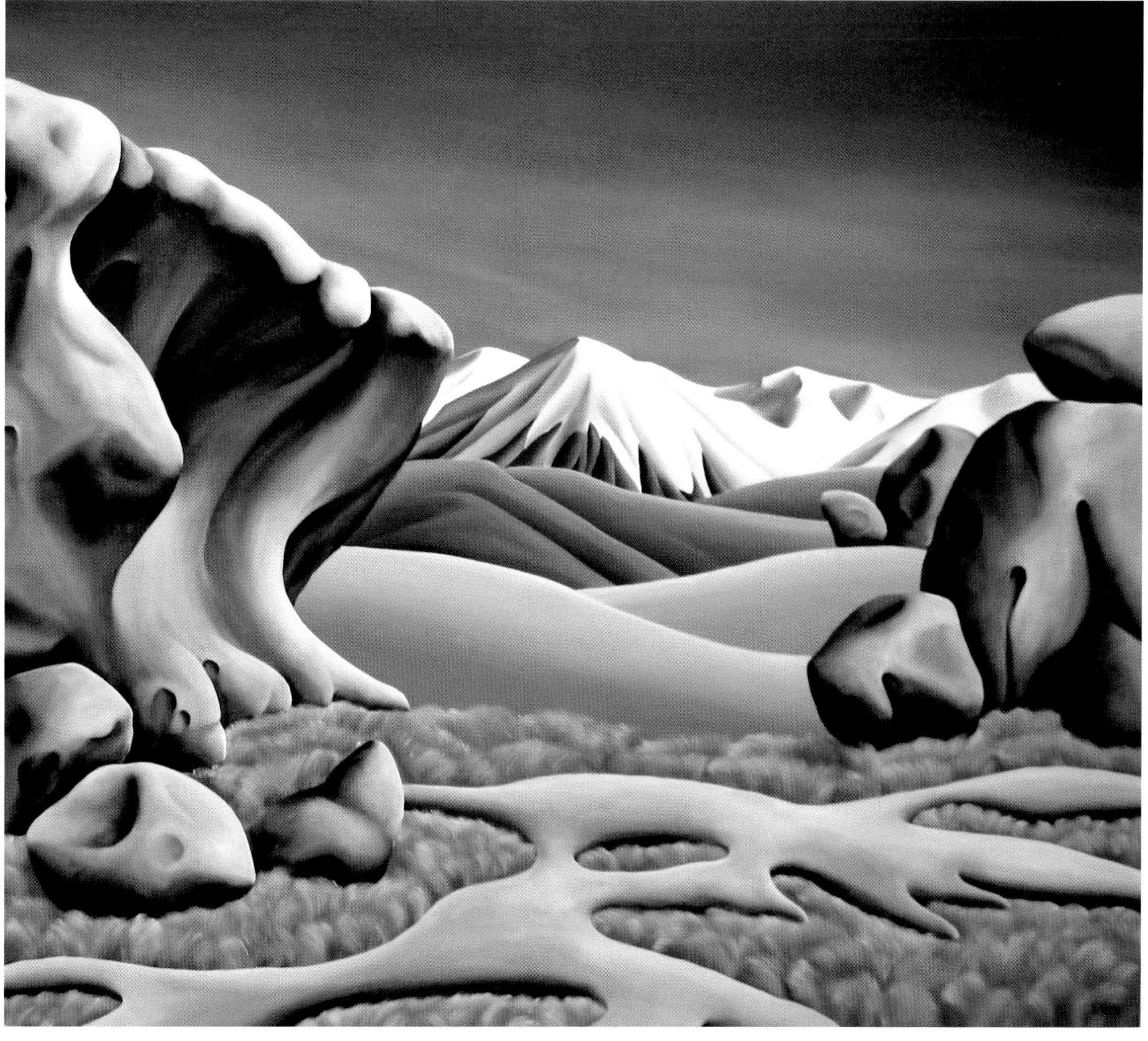

Looking east across Castle Hill Basin from where State Highway 73 crosses Parapet Rock. In the foreground Cave Stream has cut a series of incised meanders through terraces made up of gravel washed out from glacial moraine. Beyond Cave Stream is the limestone scarp of Prebble Hill, and on the left is Flock Hill. In the distance is the Torlesse Range.

Climbing out of the last of these gullies, the road travels north over sandstone and gravel terraces. Before you leave the basin it is worth pausing at Parapet Rock to look down the Cave Stream Valley at the magnificent view across the limestone formations to the Torlesse Range. Immediately below, the small stream follows a series of meanders cut at a time when its flow was much more powerful. Parapet Rock follows the edge of the limestone of Flock Hill back across Cave Stream.

Lake Pearson is just north of Flock Hill as the SH73 enters the Cass Basin. It formed in a hollow carved by ice and is bordered on the east side by the glacially steepened side of Purple Hill, and on the west by large tussock-covered shingle fans.

Cass Basin

Ice-eroded hills, valley floor with glacial lakes, moraine deposits and alluvial fans

The drive down the Craigieburn cutting from Castle Hill leads into another area surrounded by mountains. Landforms in Cass Basin were strongly modified by the Waimakariri Glacier. Its floor is quite extensively covered with moraines left behind after the ice retreated. More recently, the surface has been shaped by large gravel fans. Cass Basin also contains several glacial lakes.

Towards the northern end of the basin there is more dramatic evidence of ice erosion. On the east side, behind the Cass railway depot, Sugarloaf Hill has been trimmed and steepened, and directly north the road reaches the Waimakariri River through a gap whose irregular rocky surface was ground down by ice extending into the Cass Valley.

Ice has also shaped two rocky mounds at the northern end of the basin into classic roches moutonnées (literally, 'fleecy rocks') – smoothed on the upstream side and plucked and scoured into rocky outcrops on the downstream side.

The highway crosses a large shingle fan that forms the edge of Lake Grassmere to the east. Its other side is bounded by glacial moraines extending down the Craigieburn Valley, which is followed by the railway line and a side road off the highway. Down this valley are a series of further, smaller lakes – Sarah, Hawdon and Marymere – filling ice-excavated hollows partially blocked by moraine.

Right: Looking north to Cass over shingle fans. Lake Grassmere is to the right. The roches moutonnées in the distance are clear evidence of the work of ice in scraping and plucking the surface of the hills.

Below: A painting by Sydney Lough Thompson of Lake Sarah, looking north towards Cass and the Waimakariri Valley behind the ridge.

Private collection, courtesy A. Thompson

View up the braided Waimakariri River, looking into its headwaters in the Main Divide, from the Mount White road.

Middle Waimakariri Valley

Large, glacially shaped valley, shingle riverbed, downstream glacial deposits and alluvial terraces

Leaving the Cass Basin, the West Coast road travels over a glacially scraped spur and enters the spectacular middle Waimakariri Valley. Some two kilometres across, with mountains on each side evenly trimmed and steepened by ice, the valley floor consists largely of grey shingle divided by intersecting streams, with alluvial grass-covered flats along the northern side.

Terraces cut through outwash gravels on north side of the Waimakariri along the Mount White road.

It is hard to imagine the sheer size of the glacier that filled this valley to the height of the shoulders on the bordering mountains, and gouged the floor perhaps hundreds of metres below the present shingle level. Upstream, the valley stretches almost 20 kilometres beyond the Bealey confluence into the Main Divide. Downstream, it extends a similar distance as far as the junction with the Esk River.

The best way to see the landforms of the downstream valley is to turn down the Mount White road, cross the river, and drive for about 10 kilometres, skirting the flanks of Mt Binser. From this point magnificent views of the extensive riverbed and surrounding ice-trimmed mountains can be enjoyed. A short distance further on the north bank a suite of terraces have been created by successive levels cutting into deposits of outwash gravels from the end of the glacier.

Beyond Mount White Station large deposits of outwash gravels left by the glacier have been eroded and deposited further by water action. Unlike the huge Rakaia Glacier, which reached the Canterbury Plains, its Waimakariri counterpart extended as a relatively narrow finger into the gorge. Lake Letitia, near the end of the road, occupies a glacially carved trough in the greywacke hills.

Alpine ranges

Glacial-erosion features – glaciers, arêtes, cirques, oversteepened valleys, moraine

The headwaters of the northern tributaries of the Waimakariri – rivers such as the Poulter, Hawden, Mingha, Bealey, Crow and White and, of course, the Waimakariri itself – are well known to trampers. To reach many of the mountains of the Main Divide you have to be prepared for some solid days toiling up the valleys. Fortunately, those less fit can get right into the heart of some of the more spectacular examples of ice-sculpted alpine landforms by taking the road over Arthur's Pass.

From near the top of the pass, you can look west to the upper slopes of Mt Rolleston (2,271 metres). This and the other higher peaks to the west, including Mt Murchison (2,400 metres), have small glaciers clinging to their summit slopes. The largest of them – 700 metres in length – is the Marmaduke Dixon, at the head of the White River. These glaciers are the tiny remnants of the huge ice sheets that shaped almost all of the surrounding landforms. Valleys such as the Bealey have been gouged into U-shaped troughs, and the rock faces above the treeline are generally very steep and scraped and smoothed by ice. Often the height of the former glacier is marked by a shoulder where a ridge suddenly steepens before dropping to the valley floor. High up there are several examples of armchair-shaped cirques (for example, Avalanche Peak) that have been excoriated by the plucking and

A small glacier on the upper slopes of Mt Rolleston.

Temple Basin, an example of a cirque and hanging valley.

Irregular dumps of moraine near the top of Arthur's Pass.

scraping action of a side glacier. The jagged ridges, or arêtes, are the result of ice that has plucked at both sides by a long process of freezing and thawing.

Near the top of the Arthur's Pass saddle, the valley floor is covered with stony ridges marking the floor of a glacier that once flowed right across the pass. Numerous small tarns can be found in ice-scoured hollows, and piles of lateral moraine abound. The well-signposted Dobson Walk, near the summit, provides an excellent means of viewing glacial landforms as well as alpine vegetation.

The alpine features so splendidly illustrated at Arthur's Pass are typical of the headwaters of all the surrounding alpine tributaries of the Waimakariri, and also the sources of the Rakaia and Rangitata Rivers. The whole area along the Southern Alps is largely shaped by the eroding action of the rocks embedded in moving ice, and even lower down in the valleys there is evidence of ridges being ground down to an irregular rocky terrace (such as near the junction of the Bealey and Waimakariri Rivers) and of rock surfaces scraped and smoothed by rocks that have been frozen in the ice and have acted as prehistoric bulldozers. The results of this process can be seen in the vicinity of the Bealey Hotel.

Landforms of the Rakaia Valley

The roads on each side of the Rakaia River lead to a huge, glacier-shaped valley similar to the Waimakariri, but the approach is very different. Because there was no mountain barrier, the Rakaia Glacier spilled directly onto the Canterbury Plains. On the north side of the river this extensive ice sheet left a distinctive legacy of glacially carved lakes, moraine dumps and outwash gravel terraces.

Looking up the Rakaia Gorge from below the twin bridges on SH72.

Rakaia Gorge

Gorge cut through gravels and underlying rock, shingle fans, terraces, outwash plains

Shortly after passing through Windwhistle when travelling west on SH72, you suddenly arrive at the lip of a gorge, with the wide expanse of the Rakaia River over a hundred metres below.

At this point it is worth looking across the flat terraces above the gorge, stretching from Mt Hutt on the south side, back to the Malvern Hills eight kilometres to the north, and visualising this expanse covered with ice. At the greatest extent of the Rakaia Glacier, its tongue extended onto the edge of the Canterbury Plains.

After each ice retreat, copious quantities of shingle were carried down the river, gradually forming the vast fan that became the central part of the plains. The gravel built up to the height of the imposing terraces on either side of the present gorge.

This happened on at least three occasions after the greatest ice advance, and each time further terraces were formed as the river cut more deeply. Eventually the power of the river was such that it created a gorge, exposing the volcanic rocks that formed in early Cretaceous times. Two bridges that span the Rakaia at this point are located on these rocks, where the river has cut narrow, steep-sided notches.

The steep terraces forming the north bank below Bayfields Station, the property immediately upstream from the gorge, demonstrate the extent and depth of shingle that has been carried down the river from the ice-eroded mountains. There are also cliffs of yellow loess – fine-grained sand produced from rocks pulverised during glacial erosion and transported by the fierce winds that blow down the gorge. The Rakaia Walkway provides views of the dramatic landforms of the gorge and of the coal seams and volcanic rocks that underlie the gravel deposits.

Fighting Hill to Lake Coleridge

Glacial deposits – terminal, lateral and medial moraines, medial moraines, kettle holes, tarns, outwash stream channels, kames and kame terraces

On the road up the north side of the Rakaia Valley from Windwhistle towards Lake Coleridge and the

The Rakaia River has cut deeply through the gravel and loess terraces above the gorge bridges on State Highway 72.
Courtesy Back Country New Zealand

power station village, the views to the left, across the river to the Mt Hutt Range, are spectacular. Although the landscape on the right seems to be just a jumble of hills and valleys, it is also worth examining. This area includes distinctive examples of moraine deposits left from the retreat of at least four glacial advances, and of landforms carved in the moraine by meltwater streams. Some of these are complicated and difficult to interpret.

To the north of the Snowdon turn-off is Fighting Hill, the remains of a greywacke peak surrounded by deposits of glacial moraine and outwash gravels from earlier glacial advances.

Beyond Fighting Hill to the Dry Acheron Stream is a rather confused mass of rocky ridges comprising terminal and lateral moraines from the more recent glaciers, and also kame terraces, created by layers of gravel deposited by water draining from the ice.

To the right off the power-house road, just after crossing the Dry Acheron, a gravel road follows the east side of the lake, quickly rises onto moraine terraces and then plunges into a long, steep-sided gully that leads to Lake Coleridge Station. This is a particularly clear example of a glacial meltwater channel that has been carved out along the edge of the moraine and once carried meltwater from the glacier that created the lake.

Fighting Hill, near the Rakaia Gorge on the north side of the river, was so named because it is the meeting place of the area's fierce nor'wester and southerly winds.

Lake Coleridge fills a trough carved by a glacier and blocked at the downstream end by moraine. On the left of this aerial photo is Mt Oakden, and on the right is Kaka Hill with Mt Cotton behind. In the distance is the Wilberforce Valley and the upper Rakaia (left).

Courtesy Back Country New Zealand

Lake Coleridge is about a kilometre wide by 17 kilometres long. It is a classic example of a trough over-deepened by scraping from the tongue of a glacier (the Wilberforce) and, later, blocked at its outlet by moraine. The lake is now drained at the western end, into the Harper River, although much of the water is diverted through tunnels to drive the Lake Coleridge power station, from where it flows into the Rakaia River.

Lake Selfe, on the Harper road, is a popular haunt of trout fishermen. In the distance is Mt Hennah (1,109 metres), one of the steep-sided peaks that were once completely surrounded and trimmed by the Rakaia ice sheet.

Courtesy 4 x 4 New Zealand

The Harper road

Large-scale glacial erosion of mountains and glacial deposits in valleys – moraines, kettle holes, tarns, outwash stream channels, kames and kame terraces

The road from Lake Coleridge homestead to the Harper River travels about 25 kilometres through remote mountains and valleys that have been shaped very distinctively by ice. In fact, at the height of ice advance most of the land below perhaps 1,000 metres was covered by an ice sheet or a series of intersecting glaciers.

A curious feature of this area is a number of isolated peaks with glacier-steepened sides and rounded tops. Beyond the turn-off at Lake Coleridge Station, the peak on the left is Kaka Hill, with Mt Georgina to the right. At the far end of the road, the more massive peaks of Mt Cotton and Mt Ida dominate.

An interesting feature to look for on the slopes of Carriage Drive, opposite the Lake Ida turn-off, is a clear faultline fracture that formed in geologically recent times – since the last melting of the glaciers about 13,000 years ago.

Much of the lower country is covered with rough, irregular mounds of moraine and kame terraces – flatter areas of glacial gravels deposited by streams. In some places gravel fans have been built up by streams; for example, where the Ryton River flows into Lake Coleridge.

The Harper road passes a series of tarns that have formed in hollows carved out by the glaciers, or resulting from moraine that has blocked the hollows. Georgina is the first of these, and further on are several narrow lakes – Evelyn, Selfe and Henrietta. A short distance off the main road is Lake Ida, nestling in a narrow valley and used for ice skating when conditions are suitable. Further from the main road is Lake Catherine, another narrow tarn in the valley between Mt Ida and Mt Olympus (2,097 metres).

The 'dry weather' road north from Coleridge to Lake Lyndon passes through a maze of glacial debris. After about four kilometres this road rises across terraces and moraine, and then winds across a very large glacier-margin channel to the southeast. Several other meltwater gullies in the hills to the east are not easily seen from the road.

Rakaia River and its major tributaries

Wide, straight, glacially shaped valleys, braided streams, shingle fans, alpine peaks in headwaters

The roads up either side of the Rakaia provide expansive views of this huge valley, with its wide shingle bed. On the north side of the river, the road that goes beyond Lake Coleridge village has particularly impressive views from a bluff just before Mount Oakden Station. From here one can see that the Rakaia and Mathias Rivers extend their broad valleys right up to the Southern Alps. The three other large tributaries that feed into the Rakaia from the northwest – the Wilberforce, Harper and

Looking into the upper Rakaia from a bluff near Mount Oakden Station. In the distance to the right is the Mathias River, and in the centre stands the impressive roche moutonnée of Double Hill.

Looking up the Rakaia from the Double Hill road. The width of this valley and the extent of its shingle filling is testament to the enormous size of the Rakaia Glacier that shaped it.

Avoca – have similarly impressive valleys. To see the headwaters of these rivers you have to leave the car and be prepared to tramp upstream for at least 25 kilometres. Each of these great valleys is the product of a glacier that has carved out a wide, straight-sided path, with a broad floor now covered with vast quantities of shingle through which intersecting streams form a complicated braided pattern.

On the south side of the Rakaia, the narrow Double Hill road runs hard against the steep slopes of the Mt Hutt and Black Hill Ranges, frequently crossing the lower parts of large scree fans that spill out into the riverbed.

Double Hill is a prominent roche moutonnée – the remnant of a greywacke ridge that has been partially worn down by ice – surrounded by a shingle sea.

Landforms of the Rangitata Valley

The upper Rangitata is just as majestic as the other two major high-country valleys but much less well known. As with the Waimakariri, the glacier that carved out the course of the Rangitata River was barred from access to the plains by front ranges. Access to the inland area, and to the Lake Heron Basin, is via one of two secondary roads.

The front ranges

Highlands steeply and irregularly dissected by streams

The road from Mount Somers village following the south branch of the Ashburton River through the front ranges offers the chance to see how steeply and irregularly dissected by rivers these highlands are. An interesting landform feature in the Ashburton Gorge is the small outcrops of limestone among the greywacke mountains. These are remnants of the marine sediments deposited above greywacke and exposed in a few downfaulted areas.

On the south side of the Rangitata, a road from Peel Forest passes close to where the river emerges from the gorge through the hills. At this point, on either side of the river, there are large and distinctive suites of terraces. These have been successively carved by the river through massive quantities of outwash gravels generated by glacial action in the upper reaches of the Rangitata, and later carried out to the plains.

The road climbs for about six kilometres and winds through heavily dissected front ranges. When it suddenly emerges at the middle Rangitata Valley, a view opens up that illustrates contrasting landform patterns caused by glacial erosion.

Rangitata Valley

Large, glacially widened and straightened valley, shingle bed, headwater streams from the Main Divide

The size of the glacier that carved out the Rangitata River confounds imagination. From the first viewpoint on the south-bank road, the valley extends upstream for over 30 kilometres before it divides into large tributaries that extend back into the Southern Alps.

For much of this distance the valley floor is over five kilometres across, covered on the southern side by grassy flats and on the north side by shingle and braided streams. On each side the high valley sides have been neatly trimmed by ice. On the steep hills on both sides the evidence of ice scraping remains as a series of long parallel terraces running high along the valley side.

The Mesopotamia road, on the south side of the river, cuts through terraces and crosses fans created by Bush Stream. Further up, on the north side, the Potts River also contributes its quota of shingle. An arm of the Rangitata Glacier must have once flowed northeast through the gap followed by the Potts to dump large deposits of outwash gravel in the basin around the Lake Clearwater area.

The bulk of the shingle that fills the Rangitata Valley has been brought down from the Havelock, Clyde and Lawrence Rivers – which extend far inland beyond Erewhon Station and have their origins among the peaks of the Main Divide. Those able to tramp up these valleys will be rewarded with stunning views of peaks such as D'Archiac (2,875 metres), some of which still have small glaciers, and an abundance of landforms such as cirques and arêtes similar to those described for the Arthur's Pass area.

Above: The south side of the Rangitata upstream from Mesopotamia Station is a classic example of a river shaped by glaciation. The spur has been truncated by glacial action in the process of straightening the sides of the river. Below it are parallel terraces carved through outwash gravels.

Below: Looking up the Rangitata from the road on the south side. The wide U-shaped valley and the river flats made up of enormous quantities of shingle are characteristic of this glacially modified landscape.

Above: In the middle ground of this view up the Rangitata, north of Mount Potts Station, is Mt Sunday. This splendid example of a roche moutonnée recently gained international fame as the site of the village of Eldora in the first film of *The Lord of the Rings* trilogy. In the distance is the Havelock Valley and the towering peak of Mt D'Archiac.

Below: Erewhon Station homestead sheltered behind a series of glacially carved rocky mounds known as the Jumped Up Downs. Behind these the Clyde River joins the Rangitata. The large gully on the far side of the Rangitata was the background for the setting of Helms Deep in the *Lord of the Rings* films.

Courtesy 4 x 4 New Zealand

The basins around Lakes Heron and Clearwater

Large downfaulted basins with extensive moraine and outwash gravel plains and lakes

Between the Rakaia and Rangitata Valleys are the large intermontane basins centred on Lakes Clearwater and Heron. Access to this area is by the road from Mount Somers up the Ashburton Gorge, emerging at Hakatere. Here the road forks, with a branch leading into each basin.

The road to the west travels through an open valley for about 20 kilometres before reaching the north side of the Rangitata by Mount Potts Station. The floor of much of this basin has been shaped by the outwash gravel that spilled into the valley from the Rangitata Glacier. Among the gravel mounds and terraces are several lakes that have formed in the hollows. These include Emma, Camp and the most attractive, Clearwater, a popular angling spot.

The north road from Hakatere quite quickly leads to a large and rather barren plain, where Lake Heron formed behind moraine material.

The gravel that covers the floor of this basin was produced by a branch of the Rakaia Glacier that spilled south through a gap in the hills to enter the

Above: A view across the Rangitata from the south side into the Clyde Valley. Erewhon is on the right.

Courtesy 4 x 4 New Zealand

Below: Clearwater in midwinter with ice forming around its edge. This is one of half a dozen small lakes formed among the moraine deposits left in the Lake Heron/Clearwater Basins. The Two Thumbs Range and Mt D'Archiac are in the far distance to the left.

Scree, the ever-present drapery of the high-country ranges

Anyone who has ever spent time climbing in the ranges between the Rakaia and Waimakariri Rivers, as I did in my late teenage years, will be familiar with the seemingly endless scree that covers most of the higher country in this region with a grey cap and dribbles in long rivers down the mountain sides. The angular, loose shingle provides precarious footing on steeper slopes, as many trampers have found to their cost.

But, as a teenager employed at Lake Coleridge Station for the shearing, I had cause to be grateful for the presence of one of these shingle slides. Foolishly, as young men do, I had taken up a challenge, made about 7 p.m., to tackle Red Hill (1,639 metres high and a climb of over 1,000 metres) that evening. Several exhausting hours later I reached the top with just enough light left to identify the summit marker. I was delighted to see a thin strand of shingle stretching straight down a steep face and disappearing in the darkness near the base of the mountain. A quick and reckless slide down this saved me from the prospect of a cold and unpleasant night high on an exposed alpine ridge.

Scrambling over a shingle slide in the 1950s. The steep angle of repose (about 30 degrees) of the loose, angular rocks makes scree slopes extremely unstable and hazardous to cross.

This photo illustrates how extensive the cover of scree is on the higher slopes of the high-country ranges.

Courtesy Back Country New Zealand

An aerial view looking southeast across the outwash gravels of the lower Lake Heron Basin towards Lake Emily.
Courtesy Back Country New Zealand

Looking south along Lake Stream into the upper Lake Heron Basin, with the lake itself in the distance.
Courtesy Back Country New Zealand

present basin. Lake Heron now drains northward via Lake Stream back through this gap into the Rakaia.

The glacier eroded the mountains to the north, trimmed peaks such as Sugarloaf Hill alongside the lake and extended over most of the basin. The mountains to the northeast – the Taylor and Mt Somers Ranges – are quite large and dissected by rivers, but do not show extensive evidence of glaciation. However, on the western side the Arrowsmith Range was heavily carved by ice, forming impressive peaks such as Mt Arrowsmith itself (2,795 metres), and the lower slopes are covered with glacial material.

Lake Heron formed behind the deposits of moraine to the far right. The ranges in the distance show the features of dissection by rivers.

CHAPTER TWO

Wind and Snow

High-country climate

'I like this country very well, nothing but grass and wind,' wrote Andrew Rutherford in a letter to his mother in Adelaide in 1860, describing his impressions of inland Canterbury.

How apt that description was then, and even today the northwestly wind dominates any discussion of the high-country climate. Frequently relentless, occasionally even terrifying, it is as much a part of the high country as mountains and tussocks.

The nor'wester is not the only significant element in the climate of central Canterbury, or even the most important, but it is certainly the feature that distinguishes this region from other mountain areas of New Zealand.

While a fierce wind in the valleys and basins of the Waimakariri and Rangitata, the 'old man' nor'wester reaches its peak through the Rakiaia Gorge. This is because, while the other rivers pass through ranges of hills to reach the plains, the Rakaia does not, and so its lower gorge acts as a wind funnel.

This nor'wester creating waves on Lake Coleridge heralds a storm brewing at the head of the lake and bringing heavy rain and rising water in the Wilberforce River and the other tributaries of the Rakaia.

The infamous Rakaia nor'wester

Everyone who has lived in Canterbury will be familiar with the luminous arch that regularly appears over the Southern Alps, signalling the imminent arrival of a weather system from the west. Streaky clouds stretch across the alps, leaving a gap of clear, light sky outlining the mountain ranges. High-country people are only too aware that the nor'west arch is the precursor of strong winds roaring down the valleys of the Waimakariri, Rakaia and Rangitata Rivers, with heavy rain falling in the headwaters. Cantabrians in general recognise the arch as signalling the arrival of hot, dry winds that make conditions unpleasant right across the plains.

A typical nor'wester sequence begins most commonly in spring with the gradual appearance of cloud behind the main ranges of the Southern Alps. This may build up over the mountains for several days, but the headwaters of the rivers on the eastern side will be obscured by a dense mass of grey cloud, and rain squalls will move quickly down the wide valleys of the Rakaia and its large tributaries, the Mathias and Wilberforce Rivers. Heavy rain falls in the mountains and frequently accompanies the strong winds as far down the valleys as Mount Algidus Station, where the Wilberforce meets the Rakaia.

By the time the storm reaches Lake Coleridge village the rain has usually ceased, but the wind, which has been increasingly built up by the funnelling

effects of the Rakaia Valley, has become a gale. As this sweeps down the shingle flats of the riverbed, it whips up huge dust clouds, at times largely obscuring Mt Hutt across the river. To the north, the nor'wester builds up in a similar way down the Wilberforce until the steep-sided lump of Mt Oakden parts this flow, the blasts on the south side adding to the force of the gale in the Rakaia. On the other side, it blows directly down Lake Coleridge, creating metre-high waves, then sweeps over moraine hills and terraces into the High Peak Valley and eventually over the Malvern Hills onto the Canterbury Plains.

The blast of the nor'wester reaches its climax around the terraces above the Rakaia Gorge bridge and is at full force as it bursts out onto the plains. In the biggest storms, peak gusts have been recorded at 160 kilometres an hour, with sustained winds around 115 km/h. These winds are not only ferociously strong,

Late-afternoon nor'west arch viewed from Christchurch Airport looking across the Canterbury Plains towards the Mt Hutt and Torlesse Ranges.

No wonder the early settlers were in awe of the nor'wester. It has even pushed the frame of this Ashley Smith painting out of shape!
Courtesy Ashley Smith and the Nor'Wester Café, Amberley

A nor'wester whips up dust in the bed of the Rakaia River.
Nick Groves

but they are also hot and extremely dry, with very low humidity, and sometimes these conditions last for four or five days. It is no wonder that the combination of the destructive force of the wind, the swirling dust and grit, and the heat and dryness have given the nor'wester the reputation for driving people crazy. It is even claimed that during a prolonged nor'west cycle suicide rates and incidents involving violence in Canterbury increase. One theory has it that this may be related to the electrostatic effects of friction from very dry air blowing across the land creating an imbalance of negative ions. Although such föhn winds are most prevalent in spring, they regularly reappear throughout the summer.

Eventually this weather pattern moves on and is typically replaced by southerly winds, which clear the air and bring cooler conditions and often rain. On the north side of the lower Rakaia Gorge is Fighting Hill, so named by the early settlers because this is where the nor'wester clashes with southerlies sweeping in from the plains. Another appropriate name is that of the small settlement of Windwhistle, on State Highway 72 near the gorge.

Although a strong nor'wester lasting several days remains a memorable and uncomfortable experience, it is generally not as devastating as it was for the first high-country settlers. The effects of the strong winds have been considerably reduced at ground level by the gradual growth of trees that now surround most homesteads, and by the rows of shelterbelts planted at right angles to the wind direction, particularly on the terraces on the north side of the Rakaia Valley.

The following quotation from Rosemary Britten's *Lake Coleridge* is a reminder of how disruptive the winds were before trees were planted. This describes conditions in 1911 when work began on the exposed terrace site of the power station:

Not a great day for a round of golf at Terrace Downs Resort with this gale blowing down the Rakaia Gorge.

Explaining the nor'wester

The weather system that gives rise to Canterbury's infamous wind is perhaps the most recognisable pattern on New Zealand weather maps. The north-east-bending kink in the isobars across the Southern Alps indicates that pressure on the eastern side has become much lower, causing the wind to blow more strongly into this area. The closeness of the isobars indicates the strength of the wind.

When the weather system approaches the South Island from the west, the air is forced to rise and cool (because of expansion) as it hits the Southern Alps. The water vapour that has been picked up over the Tasman Sea also cools and condenses, forming heavy clouds and bringing rain to the West Coast. These conditions spill over the Main Divide and into the headwaters of the big Canterbury rivers.

As the air descends on the eastern side, it loses the last of its moisture and begins to heat up as it is compressed. The combination of hot, dry air and lower pressure results in a föhn wind, a phenomenon known in other parts of the world where similar conditions occur; for example, east of the Rocky Mountains in North America and in valleys on the leeward side of the European Alps.

The force of the Canterbury nor'wester is considerably increased by being directed down the long, glacially shaped funnels of the big rivers and their tributaries, from which the wind emerges onto the Canterbury Plains.

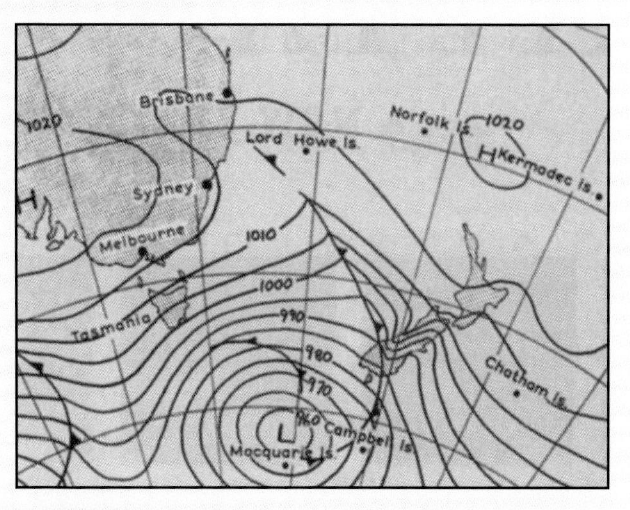

A metrological map showing a typical nor'westerly weather pattern. Note the kink in the closely spaced isobars along the Southern Alps. This indicates much lower pressure to the east and strong föhn winds blowing across the Alps into the low-pressure zone.

Why are these conditions more common in spring? To explain this requires a look at the Earth as a whole. In general, wind is caused by temperature differences between the poles and the equator. Spring is the time in New Zealand's part of the world when these contrasts are at their greatest. This causes winds to blow out from the polar area and be deflected east by the rotation of the Earth. Thus westerly gales in the southern hemisphere are at their strongest at this time of the year.

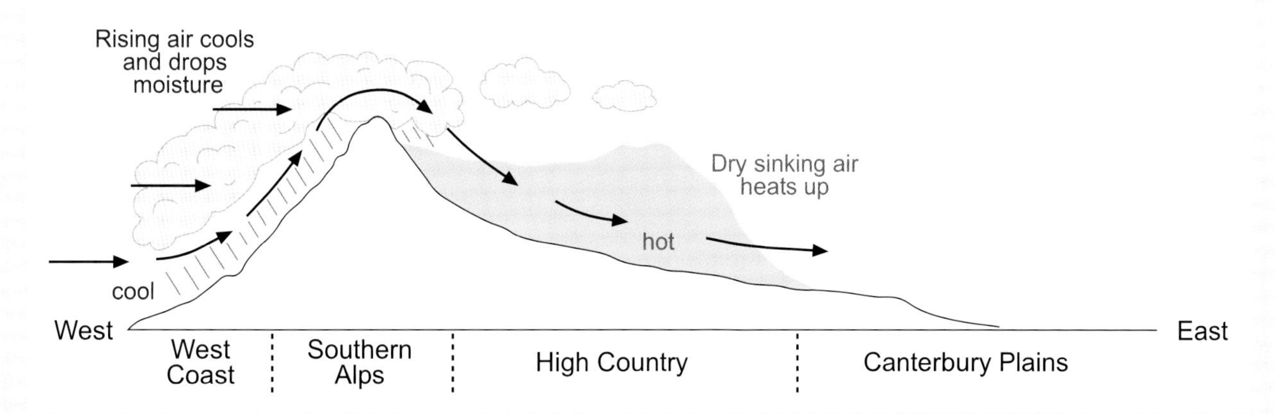

This cross-section across the South Island from Hokitika to Christchurch illustrates the processes involved in creating the nor'wester.

Diagram by Tim Nolan

> Early settlers wrote of dust blowing in clouds down the river bed making it look as though the river were on fire, with thick smoke rising from it. When the builders of the power station began work near the river their days were a battle against the wind. Dust and gravel was blown so forcibly against the unprotected building that men could hardly make themselves heard and it was often impossible to carry on . . . people told of having to crawl on hands and knees because they couldn't stand against the wind. A three year old son of Mr Kissel, the first engineer, was caught up by the wind and flattened against a wire fence, unhurt but quite unable to move.

The writings of the earliest pioneers in Canterbury reveal the enormous impression made on them by the nor'wester and the problems it caused them. The most vivid of their descriptions of this wind are from explorers and settlers in the Rakaia Gorge area, including Lady Barker, who for a few years helped her husband establish a sheep run in the Malvern Hills, in the direct path of nor'westers coming from the gorge just a few kilometres away. In *Station Amusements in New Zealand* (1873) she wrote:

> For a week past, a furious north-westerly gale had been blowing down the gorges of the Rakaia and the Selwyn, as if it had come out of a funnel, and sweeping across the great shelterless plains with irresistible force. We had been close prisoners to the house all those days, dreading to open a door to go out for wood or water lest a terrific blast should rush in and whip the light shingle roof off . . .
>
> . . . The yellow tussocks were bending all one way, perfectly flat to the ground, and the shingle on the gravel walk outside rattled like hail against the low latticed windows. The uproar from the gale was indescribable, and the little fragile house swayed and shook as the furious gusts hurled themselves against it. Inside its shelter, the pictures were blowing out from the walls, until I expected them to be shaken off their hooks even in those walls which had plank walls lined with papered canvas; whilst in the kitchen, store room, etc. whose sides were made of cob, the dust blew in fine clouds from the pulverised walls, penetrating even the dairy, and settling half an inch thick on my precious cream. At last, when our skin felt like tightly-drawn parchment and our ears and eyes had long been filled with powdered earth, the wind dropped at sunset as suddenly as it had risen five days before.

Lady Barker was not alone in her despair. Many of the other early settlers in this area who kept diaries or wrote reminiscences describe their early experiences of the nor'westers in similar terms. Mark Stoddart arrived in Christchurch from Australia in 1851 and explored the Rakaia Valley as far as Lake Coleridge

A strong, hot and dry Canterbury nor'wester makes shearing big merinos an unenviable job.

Courtesy Ashley Smith and Nor'Wester Café, Amberley

that year. He then took up a run of 8,000 hectares at Terrace Downs, on the north side of the Rakaia River just downstream from Windwhistle:

> This beautiful spot, however, has peculiar drawbacks of its own – the nor'-west winds, the curse of New Zealand, pour thro' this embrasure of the mountains with a force which must be witnessed to be believed and converts the avenue-like bed of the river into the most howling scene of desolation – horsemen are blown out of the saddle, sheep drift before it miles upon miles, cultivation is uprooted and the soil carried bodily away.

Heavy winter snow

The other dominating weather feature of the central Canterbury high country is the region's harsh winter. On the higher peaks snow remains for six months or more, and even on the floors of the basins it may sit on the ground for up to two months. Such a persistent cover is great for the Canterbury skifields, but it poses significant stock-management problems, such as the provision of winter feed and, in extreme cases, snow-raking to rescue trapped sheep.

Most significant for the economic survival of the high-country farmers are the very heavy snowfalls that take place every 10 years or so. These falls can result in heavy stock loss, at times so severe that they have forced runholders off their properties.

The majority of very heavy snowfalls are caused by the arrival from the southwest of what is known by the National Institute of Water and Atmospheric Research (NIWA) as a 'weather bomb' – a cold and rapidly deepening depression behind a strong, moist airflow. The depression forces the moist air up, resulting in large amounts of precipitation that will become snow at low temperatures. If the snow falls for several days, a metre or more may stay on the ground for several weeks.

The great snowfall of August 1867 was the first really severe episode experienced by settlers in the Canterbury high country, and was well documented. This storm particularly affected the Waimakariri area. The Enys brothers at Castle Hill noted that snow fell for 80 hours and was a metre deep around the homestead and three metres on Porters Pass. They lost over 4,000 sheep and lambs, a blow that took several years to recover from. It is notable that three neighbouring runs – Craigieburn, Riversdale and Cora Lynn – changed hands that year. The snow must have been even worse in the foothills to the east.

Snow falling on the Clent Hills homestead.
Courtesy Tussock & Beech Ecotours

Other memorable falls are recorded for 1878, 1879, 1888 and 1895 (when the Arthur's Pass road was closed for three months), 1903, 1918, 1939, 1945, 1968, 1973, 1992 and 2006. The storm in 1945 was particularly severe, with heavy snow experienced as far away as Christchurch. David McLeod, in *Down from the Tussock Ranges*, noted that conditions were

Merinos feeding on turnips grown for winter feed on Mount Arrowswmith Station in the Lake Heron Basin.

Lady Barker's ordeal in 1867

Some of my most vivid memories as a child growing up in the high country are of the 1939 snow, when our homestead was cut off by snow for three weeks from the rest of the world. This was an exciting time for children, when even a routine like going out to the woodshed to get an armful of firewood was an adventure. But this pales into insignificance beside the massive 1867 snow, the subject of a chapter in Lady Barker's *Station Life in New Zealand*. She and five others were trapped in their farm cottage for six days with virtually no food or firewood while it snowed, and were isolated for another week or so while the snow thawed. By the third day

> . . . things began to look very serious indeed: the snow covered the ground to a depth of four feet in the shallowest places, and still continued to fall steadily; the cows we knew must be in the paddock were not to be seen anywhere; the fowl house and pig styes . . . had entirely disappeared; every scrap of wood was quite covered up; both the verandas were impassable; in one the snow was six feet deep, and the only door that could be opened was the back-kitchen door, as that opened inwards; but here the snow was over the roof, so it took a good deal of work with the kitchen shovel to dig out a passage. Indoors we were approaching our last mouthful very rapidly . . . Friday, and the same state of things: a little flour had been discovered in a discarded flour bag, and we had a sort of girdle-cake and water . . . we were all more than half starved, and quite frozen: very little fire in the kitchen, and none in any other room. Of course, the constant thought was, 'Where are the sheep?' Not a sign or sound could be heard.

Eventually they managed to retrieve some frozen hens and, using fence railings, were able to light a fire and survive on chicken soup. Much of the second week, while the snow thawed, was spent on the unpleasant task of rescuing the few surviving sheep from among the great majority that had drowned or frozen to death.

exacerbated by a series of very heavy frosts, so hard that Lake Pearson was completely frozen over. He described the difficulties of snow-raking and trying to lead groups of stranded stock down to patches of partially clear areas on the sunny faces, and also of his vigils on the ridges on frosty nights, watching for kea intent upon attacking sheep trapped in the snow.

Although large numbers of sheep have perished in more recent storms, better communications and greater preparedness have diminished the threat posed by these conditions in recent years. It is also apparent that, since records began to be taken about 1860, average temperatures have been increasing and the frequency of serious snowfalls has declined, although the massive snowfall on 12 June 2006 was a powerful reminder that they may still occur.

This storm has been well recorded and analysed by NIWA scientists, who suggested that in many areas it was at least as severe as the infamous 1945 event. Heavy snow fell throughout the central Canterbury high country and was particularly deep in the area north of the Rakaia River. Over 80 centimetres was recorded at Ryton and on many of the stations in the Waimakariri catchment. This caused extreme disruption, with power lines down and homesteads isolated. In some places the snow took over two weeks to melt and supplies of hay for winter feed for stock were severely depleted.

Floods

A major side effect both of heavy rain in the headwaters of the rivers and snow melt is the regular flooding of the three great rivers and their tributaries, especially in spring. A full flood through the narrow gorges is an awesome sight. The Waimakariri, for example, may rise six metres or more as it carries an enormous volume of yellow-brown water surging through the gorge. While confined to the gorges or wide riverbeds, the floods are impressive but not usually disastrous; however, they cause considerable access problems higher up the valleys. Many high-country trampers have been trapped in huts, sometimes for days, because of rapidly rising streams.

Most affected by flooding, however, are those working on Mount Algidus, Manuka Point and, to a lesser extent, Glenthorne Stations, because the only road access to these properties is by fords across the river, the conditions of which can change in a flash and prevent crossing for weeks. The problems posed by the unstable Wiberforce River were documented by Mona Anderson, who lived at Mount Algidus for over 30 years. In *A River Rules My Life* she gives a graphic description of her return from her first time away from the station:

> Our return trip next day was most unpleasant. Jim was waiting for us in a howling nor'-west gale. 'You won't be able to see for dust in the riverbed,' he told us.
>
> He was right. The silt and fine riverbed shingle hit us in the face. I kept my eyes shut tightly, but my ears and hair seemed to be full of dust, and when I ran my tongue round my mouth I could feel grit on my teeth. We had not been home long before the rain came pelting down, and in the morning for the first time I saw the river in all its might. It was a roaring muddy-brown torrent. Waves caused by the main body of water hitting a submerged bank were leaping six feet into the air. On the surging surface of the water floated sticks and logs, and now and then a whole great tree was carried past, its branches arching slowly into the air and under again as it turned in the grip of the flood. No wonder Ron had refused when I asked for one more day in town.

A GMC truck, used to cross the Wilberforce to reach Mount Algidus Station, stuck in the flooded river in 1957.
Courtesy Corinne Crawley

A mountain climate

As in the rest of New Zealand, weather in the Canterbury high country changes regularly and quickly. It can provide a crisp, cloudless midwinter's day, with the snow-covered Taylor Range mirrored in a glassy Lake Heron; a howling nor'wester tossing the homestead trees and flattening the tussocks; or a frigid wind sweeping rain over Arthur's Pass; or a clear morning high on a range with the morning mist below creeping up the hillsides.

Putting together these varying weather conditions, we can arrive at a picture of the high-country climate. New Zealand as a whole has a mid-latitude climate, characterised by successive frontal weather systems crossing relatively narrow land masses from west to east. The special conditions in inland central Canterbury that produce nor'westers are the imposing barrier of the Southern Alps and the long, straight glacial valleys running to the east. In fact, most of the other distinctive features of the weather of the region are also attributable to the mountains and, to a lesser extent, by the inland location, away from the influence of the sea. So how does the mountain climate differ from that in nearby areas?

First, the mountains affect precipitation (rain and snow). The predominant westerly flow shows up as a very marked gradation in the amounts of precipitation from west to east. Along the Main Divide, rainfall is very heavy year round, with as much as six metres per annum in parts of the Southern Alps. Only 20 kilometres east of the mountains, annual rainfall declines to as little as a metre in places like Castle Hill or Lake Coleridge. Further east again, the total increases slightly in the hill country abutting the Canterbury Plains, reflecting the additional precipitation that comes in from the south. On the coast, Christchurch averages about 700 millimetres of rain each year.

Second, the mountains affect temperatures. Anyone who has tramped or skied in the high country is well aware that increasing altitude means decreasing temperatures. Mountain weather forecasts regularly indicate 'freezing level at 1,000 metres', or 'snow to 1,500 metres', and it is said that temperatures drop by about one degree for every 200-metre increase in altitude. But even on the floors of the basins and valleys the mean annual temperatures are generally at least three degrees lower than in Christchurch. Altitude is again a major factor – even the lowest parts of this region, near the eastern edge of the high country, are generally close to 400 metres above sea level, and the inland valleys and basins are often 600 metres or higher.

Third, the inland location produces 'continental' effects. The sea tends to have an ameliorating effect on the extremes of temperature on the nearby land, and this region is at least 50 kilometres from the coast.

Temperatures tend to rise higher in summer and drop lower in winter than in Christchurch, which has daily and seasonal temperature ranges of around 9 or 10 degrees. In some of the inland basins these ranges are 12 degrees or even higher.

These, then, are the main factors that make the climate of the central Canterbury high country distinctive. But within this region there are variations in climate, and the following pages examine differing conditions in several areas. It should be noted that weather-recording stations are not common in the high country, so accurate data is hard to acquire. Although about 30 sites have long-term rainfall records, few other statistics been recorded except for scattered samples.

The wet, cold climate of the Main Divide

From the summit ridges of the Southern Alps, extending eastward for about 13–18 kilometres, stretches a long, narrow strip of alpine country with a distinctive climate, frequently very cold and bleak conditions with heavy falls of rain and snow. Travellers pausing at the top of Arthur's Pass to admire the view or photograph kea may be lucky enough to be greeted by fine weather, but more often will experience a biting wind, often accompanied by mist and rain, and even snow in winter.

This climatic zone is frequented only by trampers and skiers, with the only site of permanent habitation being Arthur's Pass village, at an altitude of about 740 metres. Although records are sparse, it is obvious that conditions here are often inhospitable.

The first snow of the winter at Mesopotamia: looking across the upper Rangitata to Erewhon and the Clyde Valley.
Courtesy 4 x 4 New Zealand

The very high level of precipitation is almost entirely spillover from the rain-bearing westerly winds rising over the Main Divide. Arthur's Pass village receives some four metres of precipitation per year, and rain can be expected on about 160 days, with snow falling on more than 20 of these and staying on the ground for 40 to 50 days. Rainfalls are usually heavy, averaging as much as 50 millimetres per rain day. Not surprisingly, the skies are often cloudy and the village enjoys fewer than 1,500 hours of sunshine a year. At Bealey, just 10 kilometres to the east of Arthur's Pass, the annual precipitation drops to about one and half metres.

The headwaters of the tributaries of the Rakaia and Rangitata show the same patterns of heavy rainfall, with totals dropping quickly to the east. At altitudes above 1,000 metres snow may fall at any time, and remain on the ground for three to five months during winter.

Temperature records also show extremes. Monthly average temperatures at Arthur's Pass village are about 14°C in January and 3°C in midwinter, and range from 30°C to below freezing. December and January are the only months that can be expected to be frost-free. To complete the chilly picture, in exposed places the winds blowing off the mountains and over the pass are generally strong and cold.

The mountains further east

The higher parts of the ranges to the east of the Main Divide have similar weather conditions but with rather less precipitation. The few records available confirm what trampers familiar with this area know well: the weather can very quickly deteriorate. In winter most precipitation falls as snow, and in the higher basins this remains on the ground for five or six months, while frosts may occur at any time.

There are no permanent human settlements above 750 metres in this region, and the ski-club huts in the Craigieburn Ranges and on Mt Olympus are occupied for only short periods.

The central basins and valleys

At least two-thirds of the inhabitants of the Canterbury high country live on the floors of the big valleys and basins. While these areas do not suffer the climatic extremes of the surrounding mountains, they are at relatively high altitudes and well cut off from the

Mist rising from the Castle Hill Basin on an autumn morning. The Torlesse Range is in the background.

moderating effects of the sea. More than other parts of New Zealand, except for Central Otago, they show continental influences on the climate.

These are typified by conditions at Lake Coleridge village, about halfway down the Rakaia Valley. A relatively low annual rainfall of about 850 millimetres is explained by the fact that the village is sufficently east that heavy rain from the nor'westers has usually almost ceased at this point, and it is too far from the Canterbury Plains for southerly rains to penetrate regularly. The annual total of about 1,800 sunshine hours is considerably more than further inland.

The village is at a somewhat lower altitude (360 metres) than most of the inland valleys, so its temperature ranges are slightly less extreme. Nevertheless, summers are usually hotter than on the coast, and harsh frosts and snowfalls are common in winter.

The highest homesteads are in the Castle Hill and Lake Heron Basins, at over 700 metres, where persistent snow cover means that the provision of winter feed for stock is essential on these runs.

Overall, the inland high country experiences a wide variety of weather, ranging from hot summer days, some calm and others battered by raging nor'westers, to winter days that can be brisk but bright after frosts, or swept by freezing storms.

The wetter eastern hill country

Those living on runs in the 'front country' on the edge of the Canterbury Plains experience a significantly different climate from the high-country basins and valleys. Such disparity is frequently encountered when driving over Porters Pass or up the Rangitata

This early-evening view from the summit of Porters Pass shows dense fog filling the Kowai Valley and extending out onto the Canterbury Plains. At the same time, the valleys on the inland side of the pass were enjoying clear conditions.

or Ashburton Gorges in cloud and drizzle, to emerge suddenly into an inland basin basking in sunshine.

This is because these hills are more open to influences of the sea to the east and to frontal weather that moves in unimpeded from the south. As a result, temperature ranges are smaller, there is more cloud, and sunshine hours are slightly lower than in the basins. Precipitation tends to be higher and snow is not quite so frequent, though heavy falls can occur when southerly fronts sweep through in winter.

CHAPTER THREE

Tussock and Matagouri

Natural ecosystems

The mountains have shaped the central Canterbury high country, but the vegetation clothing the area in dramatic colours gives life to the landscape.

Although a wide variety of mainly indigenous ecosystems are found throughout the region, it is the golden tussock grasslands and sombre green beech forests that leave the most enduring impression. Smaller and often less accessible communities in the mountains include alpine scrub, fellfield, snow tussock herbfields and scree. At lower levels, wetlands and riverbeds are distinctive habitats.

The habitats

Each of the high-country ecosystems is located in well-defined areas based on environmental conditions. Not surprisingly, altitude is a key factor in the ability of plants and animals to survive and thrive. The major systems tend to occur in zones up the mountains, and the flora and fauna that occupy each zone are determined by their ability to withstand temperature extremes, exposure to wind and sun, soil-moisture levels and the slope and stability of the ground surface.

A banded dotterel hiding among riverbed vegetation.
Courtesy Tussock & Beech Ecotours

The main ecosystems are described in the following pages, with comment on the conditions present and how the plants and animals have adapted to live in these conditions. There are dozens of species of plant and animal in each ecosystem, and it is beyond the scope of this book to describe them all. Instead, a few of the most typical species are looked at in detail.

High alpine ecosystems

The most inaccessible of the ecosystems comprises plants and animals that live above the alpine scrub and tussock herbfields, with the range of some extending to mountain peaks. Plants at this altitude tend to exist in small, scattered patches, sometimes as stunted shrubs tucked into sheltered areas around rocks, or as tiny isolated plants clinging to a relatively stable patch of scree.

These communities are found above about 1,700 metres, but nowhere are they easily accessible without undertaking a strenuous climb. The easiest sites to reach are from the top of Arthur's Pass, where tracks lead up to Temple Basin and other high ridges. High alpine ecosystems are similar throughout most of the mountains in the Canterbury high country, though in the drier east some distinctive species such as the vegetable sheep (*Raoulia exima*) are present.

It is a pity that these communities are so inaccessible, because they have evolved a particularly interesting series of adaptations to cope with a habitat that presents harsh challenges. They must be able to cope with temperatures that regularly drop well below freezing and to survive under persistent ice and snow cover. On the other hand, high daytime temperatures and ferocious winds combine to evaporate moisture from leaves. There is little soil in this environment, so mineral nutrients are in short supply and, for the few plants that grow on scree slopes, unstable conditions are a constant threat.

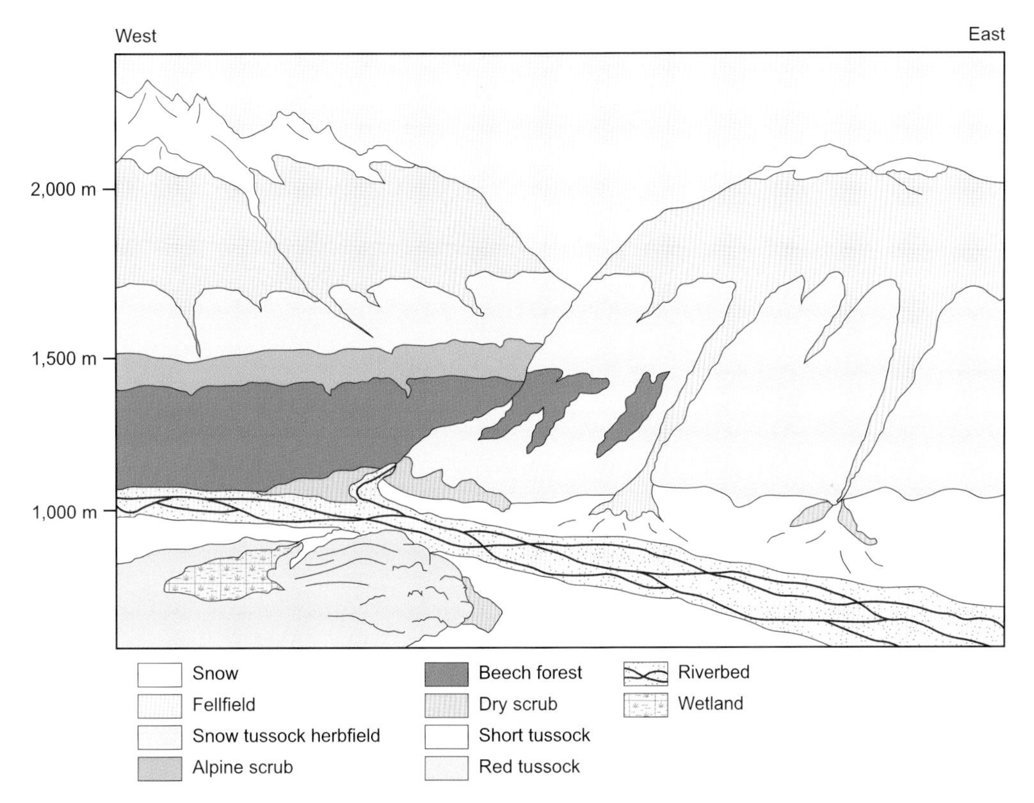

Diagrammatic summary of the zonation of Canterbury high-country ecosystems between the wetter (western) and drier (eastern) areas.
Diagram by Tim Nolan

Red tussock and matagouri are typical plants of the dry, open basins that lead to the upper Rangitata Valley.

Plants in this zone have some or all of the following adaptations:

- Their growth shape is a low-growing cushion or mat form with closely spaced branches, evolved to provide protection from the effects of wind.
- Leaves tend to be small and compressed, to reduce evaporation from their microscopic stomata. Often they have tough, impervious surfaces, and sometimes the under surface is covered with a mat of protective hairs.
- Roots are long and penetrate deeply into crevices, providing stability and reaching down to where moisture may be present.
- Internally, fluids in the cells contain a type of botanical 'antifreeze'.
- Most alpine plants have a short growing season, so flowers open rapidly after snow melt.

The following are examples of some alpine habitats and the plants they may contain.

Plants and animals living in the rocky habitat of fellfields require specialised adaptations to survive.

Courtesy Tussock & Beech Ecotours

Fellfields

These are areas of rock and scree with discontinuous patches of plants. There may be occasional small snow tussocks present, but most of the vegetation comprises small shrubs, mats of plants such as mountain daisies (*Celmisia*) with thick, hair-covered leaves, and tiny, wiry carpet grasses. Several species of whipcord hebe are found in fellfields. Their compact habit and small, tough, closely spaced leaves are typical adaptations for water conservation.

Scree slopes consist of loose angular rocks up to 15 centimetres in diameter, generally steeply sloping with an angle of repose of about 30 degrees. Most plants cannot tolerate these unstable conditions; among the few that do are the scree buttercup (*Ranunculus haastii*) and the penwiper (*Notothlaspi rosulatum*). Both put down extensive root systems, anchoring them in whatever stable, sandy soil can be found under the moving scree. Often the leafy parts can regenerate quickly if sheared off. The greyish, succulent leaves have a waxy surface that protects the plant from drying out in the heat of the sun, both direct and reflected. Willowherbs (*Epilobium*) may also be present, and rely for survival on being fast-growing and quick to produce flowers and seeds.

Among the unusual plants in the alpine zone is the vegetable sheep, which can be found on rocky ridges particularly in the Torlesse and Craigieburn

The scree buttercup (*Ranunculus haastii*) is one of a small number of plants that can survive the harsh alpine environment.

Courtesy Tussock & Beech Ecotours

The rare penwiper plant (*Notothlaspi rosulatum*) grows in partly stabilised scree up to an altitude of 1,850 metres.

Courtesy Tussock & Beech Ecotours

Whipcord hebe, with a cushion form and small, tightly spaced leaves.

Vegetable sheep (*Raoulia exima*).
Courtesy Tussock & Beech Ecotours

Ranges. The very small leaves of this plant are compacted into a grey mound to conserve moisture in dry, windy conditions.

Snow tussock herbfields

The road over Arthur's Pass traverses an example of what is perhaps the most attractive of the alpine ecosystems, the snow tussock herbfield. This is a community of tall tussocks whose narrow, flexible leaves act as a protective shield for the abundant herbaceous flowering plants interspersed among them. Together they form alpine gardens that are spectacular in the spring and summer months.

These herbfields are widespread along the main ranges of the Southern Alps above the bush and scrub and below the high alpine communities, but Arthur's Pass provides the only easy access to them unless one is prepared to do some tramping. On the drier eastern mountains snow tussocks predominate and the herbfields are often confined to mats of mountain daisies.

The plants of this zone are adapted to live in a habitat only slightly less harsh than high alpine conditions. They must be able to withstand extreme cold, strong winds and long periods covered by snow, and they grow best where soils are regularly wet, as in the damper hollows on the western mountains. The dominant plants are snow tussocks, which form large clumps, at times well over a metre tall. Without their protection many of the flowering herbs would struggle to survive. The following are some of the more spectacular plants of the alpine herbfields.

The various species of speargrass, or spaniard (*Aciphylla*), are notable for the spectacular metre-high bright yellow flowerheads they produce between November and January, and the acute sharpness of their rosettes of needlelike leaves. The latter certainly deter browsing animals, and this may be a factor in their increase in lower altitude grasslands. The larger species of speargrass are widespread in damp western herbfields but are less common in the east.

Speargrass (*Aciphylla*) in flower.

Mt Cook 'lily'/buttercup (*Ranunculus lyallii*).
Courtesy Tussock & Beech Ecotours

Several species of buttercup are common in the alpine herbfields, most being easily recognisable by their yellow flowers among the tussocks. The best known is the misnamed Mt Cook lily (*Ranunculus lyallii*), which is actually the largest buttercup in the world and, unlike smaller species, is white-flowered. They are common in local patches in the wet herbfields on the main ranges of the Southern Alps, and it is worth visiting Arthur's Pass in December or January to see these handsome plants in flower. They may grow to more than a metre tall, and individual flowers are often over 15 centimetres in diameter.

Throughout the alpine herbfields the typical white flowers of various species of mountain daisy (*Celmisia*) are prominent, and in drier areas they are generally the dominant flowering herbs, in places grouping as extensive mats. A few small species have thin, flexible leaves, but the majority form large rosettes of thick, tough but soft leaves, well suited to retaining water. Particularly showy is the large mountain daisy (*C. coriacea*), which bears flowers 10 centimetres or more in diameter.

In the wetter areas a profusion of small flowering herbs can be found low down among the tussocks. These include species of eyebright (*Euphrasia*), mountain foxglove (*Ourisia*), snow groundsels (*Senecio*) and even several orchids and native carrots (*Anisotome*). Widespread throughout the alpine grasslands are several species of tiny snow gentians, particularly *Gentiana bellidifolia*, whose white flowers are on display from February to April.

Alpine scrub

Directly above the bushline, particularly in the wetter western mountains, a narrow zone of more continuous scrub gradually merges into the more scattered high alpine vegetation. This vegetation is often quite dense, at least waist-high, and trampers know how difficult it can be to scramble through. These shrubs, typically with a dense habit, stiff, closely spaced branches and small, tough leaves, include species of *Hebe*, *Dracophyllum*, *Senecio* and *Olearia*, plus divaricating coprosma, alpine totara and mountain toatoa. In sheltered places at Arthur's Pass, mountain ribbonwood is present.

Alpine daisy (*Celmisia coriacea*).

Eyebright (*Euphrasia revoluta*).
Courtesy Tussock & Beech Ecotours

The little mountain heath *Pentachondra pumila*.

These two photos of alpine scrub at top of Arthur's Pass illustrate the variety of species and forms within this ecosystem.

Left: A dense thicket of large shrubs.

Right: A scattering of snow tussocks mingle with wiry turpentine bushes and snow totara.

The common turpentine bush (*Dracophyllum uniflorum*) is a rather drab reddish-brown shrub with long, slender leaves on tightly bunched branches.

Snow totara (*Podocarpus nivalis*) is an excellent example of a shrub adapted for alpine conditions, being much branched and with tightly spaced, leathery leaves. Unlike its close relative the mighty totara tree, it is a prostrate, spreading shrub usually under a metre tall. Its bright red fruits appear in autumn and are a good food source for alpine animals.

Alpine animals

The incredibly harsh environment of the alpine zone would seem an unlikely home for animals, yet many have found ways to survive. The most obvious is the world's only mountain parrot – the kea (*Nestor notabilis*). A roadside stop at the top of Arthur's Pass will frequently be rewarded with visits by these remarkably clever birds, which entertain with their aerobatics (enhanced by the orange coloration of their underwings) and their curiosity, particularly in investigating items of human origin. They are notorious for stripping rubber mouldings from cars and destroying unattended packs. Kea spend much of their time during the summer above the bushline, feeding on a variety of berries, leaves and buds of alpine plants.

Snow totara (*Podocarpus nivalis*).

Kea (*Nestor notabilis*) in the car park at the top of Arthur's Pass.

Kea were once very common in the high country, but because some birds developed the habit of killing sheep by tearing holes in their backs and exposing the intestines, they came to be regarded as serious pests and a bounty was put on them. It is recorded that as many as 40 birds were shot in one hunt in the Wilberforce Valley. By the 1950s kea were rapidly declining in numbers and, although now protected, they are now rarely found far east of the main ranges.

Several smaller birds such as the skylark (*Alauda arvensis*) and native pipit (*Anthus novaeseelandiae*) occasionally venture into the fellfields, but the rock wren (*Xenicus gilviventris*) is the only species that spends its whole life in the mountains. These tiny and rare birds can sometimes be seen darting around rocks and scrub, looking for insects. It may be that they survive through winter by hibernating in a dry, sheltered nook.

The wren is not the only animal that occupies the upper reaches of the high country. A number of alpine plants are insect-pollinated, and a careful look on a calm summer day will soon reveal a variety of insects in surprisingly large numbers. Various species of small moth and cricket are common, together with flies, beetles and butterflies. Most insects either hibernate or are present as eggs during winter, and are only active during the summer months.

This alpine grasshopper is one of a large variety of insects that live in the alpine ecosystems.
Courtesy Tussock & Beech Ecotours

Beech forest

In striking contrast to the shimmering gold of the tussocklands, dark green beech forests dominate the slopes of the highlands along the eastern edge of the Main Divide. In the upper headwaters of the tributaries of the Rangitata and Rakaia they are accessible only to musterers, hunters and trampers, but they are more extensive in the upper Waimakariri Valley. Here the valleys of the western tributaries such as the Andrews, Hawdon, Poulter, Bealey and Crow are clothed in forest and are readily accessible from the Arthur's Pass road.

The upper limits of the forest form a precise line along the valley sides, governed by temperature, at an altitude of up to 1,300 metres in the Waimakariri area and slightly lower further south. Rainfall has had a similar controlling role in marking the eastern edges where the influence of westerly rain is reduced.

There are remnant patches of drier beech forest further east, such as in the Craigieburn area and on the western slopes of Mt Torlesse. Along the front hills facing the Canterbury Plains the influence of rain from the south and east is sufficent to support several further patches of beech forest in areas such as Peel Forest, Mt Somers, the Harper Hills, the Big Ben Range and the eastern slopes of Mt Torlesse.

The mantle of beech trees seems drab and monotonous compared with other indigenous New Zealand forests, as it is dominated by a single genus – *Nothofagus*, the New Zealand beech. Mountain beech (*N. solandri* var. *cliffortioides*) grows at higher altitudes than the other species. It is somewhat shorter and has smaller leaves, and its own growth form changes with the colder temperatures of increasing altitude. Trees on valley floors may reach 15 metres, but at higher altitudes they have less fungus on their trunks and are much shorter – at the upper treeline they may be no more than two to three metres tall.

The amount of rainfall causes variations in the structure of the forest. A scramble through the wetter western forests near the Main Divide can be quite a challenge because of the dense undergrowth of ground ferns and shrubs, but the patches of beech forest in the drier east, such as the Craigieburn Range, tend to be more open.

The interior of the beech forests can be beguiling, with light filtering through layers of branches covered with tiny oval leaves. The shrub layer in the interior usually includes several species of small-leaved coprosma, weeping matipo (*Myrsine divaricata*) and a few larger-leaved species such as

The clear edge of the upper limit of beech forest is obvious in this view of the ranges across the river in the upper Waimakariri Valley.

broadleaf (*Griselinia littoralis*). Where there is sufficent light, there will be numerous seedling beech trees. Mosses, lichens and fungi are abundant on the trunks and forest floor in lower and damper areas, along with ferns such as prickly shield fern (*Polystichum vestitum*). Near streams a wider variety of shrubs, such as the tree fuchsia (*Fuchsia excorticata*), are present, with mountain ribbonwood (*Hoheria lyallii*) near shingle slides, and at higher altitudes subalpine shrubs begin appearing. The very showy red mistletoe (*Peraxilla tetrapetala*) has largely been eliminated by browsing possums, but one surviving and easily accessed patch at Bealey Spur can be admired in flower from spring to midsummer.

The beech trees play a crucial role in the whole forest ecosystem. Flowering occurs every three to five years, but when it does the forest is covered with a copper tinge created by the tiny reddish flowers, and a huge quantity of seed is produced. These 'mast' years, as they are called, have a dramatic effect on animal populations in the forest. Rats and mice feed on the beech nuts and multiply, attracting predatory stoats. When rodent numbers

Interior of beech forest opened up by browsing animals.

decline, mustelids turn to birds, some species of which have also benefited from the abundant seed.

Another important role played by the beech trees is in acting as host to vast numbers of a tiny scale insect that burrows into bark to feed on the sugary sap. Excess sap is secreted as drops of honeydew on the end of a tiny protruding hair. The abundant sooty mould that covers the trunks of most beech trees is dependent on the nourishment provided by this honeydew, which is also a major food source for insects, lizards and birds.

In recent years this complex web of interdependent organisms has been thrown out of balance. Sometimes in summer the forest in dry areas such as the lower Hawdon Valley and Craigieburn almost becomes a no-man's-land. Those entering these areas are greeted by the ominous-sounding humming produced by huge populations of introduced wasps (common and German), which compete for the honeydew and attack other organisms. Their prey includes many of the insects that are a food supply for birds, and they have also been known to kill newly hatched nestlings. It has been estimated that in some years South Island beech forests are home to 20 billion wasps. Fortunately, they cannot survive in the wetter forests or those above about 600 metres.

Trunks of beech trees showing drops of honeydew on the black mould.
Courtesy Tussock & Beech Ecotours

Scrublands

The dry lower margins of the beech forest host one of New Zealand's most distinctive plant associations, the communities of divaricating shrubs. These tall (up to three metres), drab, spiky bushes carry a dense mass of thin, wiry branches interlacing at right angles. Their tiny leaves are so inconspicuous the plants give the impression of being grey and leafless. They include several species of *Coprosma*, especially *C. propinqua* and *C. parviflora*. When closely spaced, these can be difficult to penetrate, especially if mixed with thorny bush lawyer or ferociously spiny matagouri. How divaricating shrubs evolved their unusual twiggy form is still being debated. Perhaps it serves to protect the leaves from adverse weather conditions, and it has also been suggested that divarication is an adaptation to reduce browsing by moa.

The scrub in this rocky gully on the road to Mt Olympus contains matagouri and a variety of other species.

Close-up of a coprosma showing its divaricating habit.

No plant better epitomises the harshness of the high country of central Canterbury than matagouri (*Discaria toumatou*), once commonly known as wild Irishman. North of the Waimakariri this is replaced by manuka, but our region is the heart of matagouri country and it would be hard to find a landscape east of the beech forest without a scattering of these distinctive shrubs poking menacingly above the tussocks on hillsides and river terraces.

Matagouri can grow to three metres high and has a more open divaricating habit than the coprosmas. Its branches are covered with ferocious spines, up to three centimetres long, which render the tiny leaves all but invisible.

Wherever conditions are particularly dry, matagouri thrives. It grows vigorously on terrace edges, in riverbeds and on the lower parts of stabilised shingle slides, to the extent that the base of an old fan is often outlined by the matagouri scrub. Extensive patches occur in the moraine country northeast of Lake Coleridge in the Rakaia Valley.

Tussock grasslands

Tussocks are the iconic plants of the Canterbury high country east of the beech forests, where they are the dominant ecosystem, covering hillsides with a tawny, rippling carpet. Although at a glance the various tussocks appear similar, they form quite complex associations that are in the process of change, both in species present and in location.

The grasslands of today are highly modified. A thousand years ago the beech forests that are now confined to western areas covered much of the high country right to the Canterbury Plains. Above the forests, snow tussock grasslands reached to about 2,000 metres, and short tussock grasslands were largely confined to river terraces and the floors of the basins and valleys in the drier areas.

Pollen analyses and charcoal deposits indicate that the arrival of Maori in the region coincided with extensive burning of the beech forest. As the trees disappeared, the snow tussock associations spread down the hills to altitudes as low as 700 metres, and the short tussock grasslands moved up to that height. After the pastoralists moved in, from about 1860, regular burning and grazing had further effects on the tussocklands. The snow tussocks were pushed further up the slopes again and were replaced by short tussocks, which withstand fire better and are less palatable to sheep. On the valley floors burning and sowing has resulted in even the short tussocks being replaced by hardy exotic grasses. In some heavily used areas these are giving way to various species of small flatweeds, especially hawkweed (*Hieracium*).

Matagouri (*Discaria toumatou*) stems and spines.

Tall tussocks

Snow tussocks form large clumps, sometimes over a metre tall, consisting of as many as 400–500 stems, each with four or five very narrow leaves. They are extremely slow-growing, but because stems are regularly replaced as they die, snow tussocks live for many years and these wiry, brown- to gold-coloured

Snow tussock (*Chionochloa*).

plants are ideally suited to the conditions on the higher slopes. Flexible stems enable these tussocks to withstand the onslaught of winds and thick coverings of snow by simply bending, while the leaves are curled into tubes to provide protection from water evaporation from stomata on the undersurfaces. The dense base of the plant collects and retains moisture.

The genus *Chionochloa* includes more than a dozen species, several of which hybridise readily, so it is often difficult to distinguish between them. Throughout the higher and wetter parts of the Canterbury high country the two most common species are the curled snow tussock (*Chionochloa crassiuscula*), particularly at higher altitudes, and the broad-leaved snow tussock (*C. flavescens*), which is more common in western habitats. In the drier areas further east, and especially at lower altitudes, the narrow-leaved snow tussock (*C. rigida*), with its large clump of reddish stems, dominates. Patches of red tussock (*C. rubra*) can be seen in some lower, more poorly drained areas, particularly in the Lake Heron Basin.

Some of the higher slopes support mats of the deep green carpet grass *Chionochloa australis*. This low-growing plant is familar to trampers, who are wary of its slipperiness when it is wet or snow-covered.

Tussocks provide good shelter for other species of grasses and herbs, though in areas where extensive grazing and burning have occurred these are being replaced by indigenous cushion species, especially celmisias, and the small blue tussock (*Poa colensoi*).

Short tussocks

Pastoral farming practices saw the short tussocks taking over most areas below 700 metres, with the most common species being the spiky-stemmed hard tussock (*Festuca novae-zelandiae*), silver tussock (*Poa cita*) and *P. colensoi*. These recover well from burning and their new shoots are very palatable to sheep. Other indigenous plants that have gained a foothold include speargrass, various celmisias and Maori onion (*Bulbinella angustifolia*), but the most significant change has been the introduction of hardy exotic grasses. Browntop and sweet vernal, in particular, now dominate the lower grasslands of the Canterbury high country.

This winter scene in the upper Rangitata Valley near Mt Sunday (in the right middle distance) shows hard tussock (*Festuca novae-zelandiae*) protruding through the snow cover.

This tussock grassland in the upper Rangitata has been modified by farming practices. Hardy species of *Celmisia* appear alongside matagouri bushes and exotic grasses.

Rare and unusual plants

This region has a number of rare and unusual plants. Most accessable are some that grow in the limestone soils around the rocks of Castle Hill. The special character of this area was recognised in the mid-1950s when a six-hectare area was fenced off as the Lance McCaskill Nature Reserve, now the site of the longest-running plant-monitoring project in New Zealand. Best known is the Castle Hill buttercup, thought to be a variant of the scree-dwelling *Ranunculus crithmifolius*, but is larger, with bigger leaves and flowers more readily, probably because it grows in a less harsh environment. Another consequence of fencing to exclude browsing animals has been the increase in numbers of several other small and very rare species that tend to grow best in limestone areas.

Other very specialised plants can be found high in the mountains, particularly in alpine-bog associations and on loose scree. A good example of the latter is *Cotula atrata*, a little black fleshy daisy that grows at altitudes between 1,250 and 2,000 metres on drier mountains such as the Torlesse Range.

Cotula atrata.
Courtesy Tussock & Beech Ecotours

Maori onion (*Bulbinella angustifolia*) is a small bulb whose bright flowerheads grow to 30–60 centimetres high in the spring and add to the attractiveness of the tussocklands. It has increased as a result of grazing and burning, particularly in the lower and damper grasslands, and is quite common around Lake Clearwater.

Animals in the tussock grasslands

The presence of animals in the grassland environment seems sparse at first glance, but closer inspection reveals that it provides a rich invertebrate life with a ready supply of food and shelter. Recent research indicates that hundreds of species live in the tussock communities: one study counted about 5,000 individual invertebrates, belonging to some 700 species, in one square metre of tall tussock grassland. It has been said that the tussock grasslands provide a habitat for the most diverse and spectacular insect fauna in any New Zealand natural ecosystem.

Most noticable are the orange tussock butterflies, a variety of crickets and cicadas, and a wide range of moths, flies and wasps. Less conspicuous are spiders, centipedes, snails and many other invertebrates.

These provide food for several species of skink and gecko, as well as those birds – especially introduced skylarks, sparrows and yellowhammers – that have successfully adapted to tussock grasslands. At the top of the food chain are Australasian harriers (*Circus approximans*), often referred to as hawks, and the occasional New Zealand falcon (*Falco novaeseelandiae*).

The New Zealand falcon (*Falco novaeseelandiae*) is the smaller of the two predator birds present in grasslands and is much less common than the Australasian harrier.

This gecko (***above***) and little butterfly (***right***) are just two of a wide range of small animals that thrive in high-country habitats.

Courtesy Tussock & Beech Ecotours

Above: A cricket perfectly camouflaged on the hawkweed *Hieracium*.

Courtesy Tussock & Beech Ecotours

Below: High-country riverbeds are mainly bare shingle and exposed to periodic flooding. Among the few plants that can survive here are mat plants such as scabweed and willowherb.

Riverbeds

The beds of the three large rivers and their tributaries are similar in appearance. These environments are very exposed, swept by wind and dust, subject to temperature extremes and occasional floods powerful enough to transport shingle, tree trunks and even large rocks.

In areas where shingle regularly moves, the only plants likely to be found are a few tiny willowherbs (*Epilobium*) or mats of scabweed (*Raoulia tenuicaulis* or *R. australis*). Where the bed is more stable, a number of other species cling to whatever soil they can find. These low-growing, tough plants include bidibid (*Acaena*), native daphne (*Pimelia prostrata*), the tiny patotara (*Leucopogon fraseri*) and wiry mats of *Coprosma brunnea* and *Muehlenbeckia axillaris*. In some places the riverbeds have been invaded by hardy exotic plants such as broom and gorse (particularly in the Rakaia), lupins (at the head of Lake Coleridge) and willows.

Given their inhospitable nature, the high-country riverbeds are home to a surprising variety of birds.

Black-backed and black-billed gulls, black-fronted terns, pied oystercatchers, pied stilts and pipits all use shingle areas in the braided rivers for nesting sites, as do banded dotterel and wrybills where they can escape the attentions of predators lurking in the encroaching vegetation.

Where currents are slow within the rivers, there is a rich insect life and a few small native bullies eke out an existence, but the main fish present are introduced trout or salmon.

Above: A wrybill (*Anarhyclius frontalis*) on a precarious riverbed nest. This bird has an extremely unusual curved beak, useful for finding food under stones.

Courtesy Tussock & Beech Ecotours

Wetlands

Wherever the land has been shaped by glaciation, lakes and swampy areas have developed in hollows gouged out by the ice. The largest body of water in the Canterbury high country is Lake Coleridge, but there are more than 20 other lakes or significant tarns, including a dozen in the Hakatere Basin, and numerous wetlands in the region.

Many of these are a focus for birdlife, in some cases species that are very rare elsewhere: for example, the

Below: The wetland around the western edge of Lake Heron is a popular habitat for waterbirds.

endangered southern crested grebe nests at Lakes Heron, Clearwater and Pearson. The Ashburton lakes are an important haven – surveys have recorded as many as 6,000 waterbirds, from at least 14 species, there. Among these are New Zealand scaup, shoveler, grey duck, grey teal, paradise shelduck, Australasian bittern, Australian coot, marsh crake, pukeko, and black and little shags, plus the introduced black swans and Canada geese. Lake Heron and the Maori Lakes are nature reserves, and Clearwater is a wildlife refuge.

Wetter parts of the grasslands are often marked by the growth of red tussock and sedges, and around the margins of some lakes a range of small moisture-loving plants (including the insectivorous sundews) can be found.

Birds of the Canterbury high country

Over 50 species of birds live in the various habitats in the central Canterbury high country. A third of these – almost all birds of the open grasslands – are introduced. The following is a checklist of all but a few very rare species.

Australasian crested grebes live only on the inland lakes of the South Island.
Courtesy Tussock & Beech Ecotours

Alpine areas
Kea (*Nestor notabilis*), rock wren (*Xenicus gilviventris*).

Beech forests
Bellbird (*Anthornis melanura*), tui (*Prosthemadera novaeseelandiae*), yellow-crowned parakeet (*Cyanoramphus auriceps*), robin (*Petroica australis*), tomtit (*Petroica macrocephala*), rifleman (*Acanthisitta chloris*), grey warbler (*Gerygone igata*), fantail (*Rhipidura fuliginosa*), silvereye (*Zosterops lateralis*), long-tailed cuckoo (*Eudynamys taitensis*), and several other quite rare species.

Open grasslands
Australasian harrier (*Circus approximans*), New Zealand falcon (*Falco novaeseelandiae*), spur-winged plover (*Vanellus miles)*, New Zealand pipit (*Anthus novaeseelandiae*).
Introduced: Blackbird (*Turdus merula*), song thrush (*Turdus philomelos*), dunnock (*Prunella modularis*), skylark (*Alauda arvensis*), starling (*Sturnus vulgaris*), sparrow (*Passer domesticus*), chaffinch (*Fringilla coelebs*), redpoll (*Carduelis flammea*), goldfinch (*Carduelis carduelis*), greenfinch (*Carduelis chloris*), yellowhammer (*Emberiza citrinella*), magpie (*Gymnorhina tibicen*), California quail (*Callipepla californica).*

Riverbeds
Black-backed gull (*Larus dominicanus*), black-billed gull (*Larus bulleri*), black-fronted tern (*Sterna albostriata*), banded dotterell (*Charadrius bicinctus*), wrybill (*Anarhynchus frontalis*), pied stilt (*Himantopus himantopus*), pied oystercatcher (*Haematopus ostralegus*).

Wetlands
Australasian crested grebe (*Podiceps cristatus*), marsh crake (*Porzana pusilla*), black shag (*Phalacrocorax carbo*), little shag (*Phalacrocorax melanoleucos*), New Zealand scaup (*Aythya novaeseelandiae*), shoveler (*Anas rhynchotis*), grey duck (*Anas superciliosa*), grey teal (*Anas gracilis*), paradise shelduck (*Tadorna variegata*), Australasian bittern (*Botaurus poiciloptilus*), Australian coot (*Fulica atra*), pukeko (*Porphyrio porphyrio*), welcome swallow (*Hirundo tahitica*).
Introduced: Black swan (*Cygnus atratus*), Canada goose (*Branta canadensis).*

Possibly present in some areas
Kingfisher (*Halcyon sancta*), morepork (*Ninox novaeseelandiae*), brown creeper (*Mohoua novaeseelandiae*), great spotted kiwi (*Apteryx haastii*), white-faced heron (*Ardea novaehollandiae*), chukor (*Alectoris chukar*).

CHAPTER FOUR

From Briar Rose to Wildling Pine

Exotic invaders

It is tempting to look upon the Canterbury high-country valleys as one of New Zealand's 'natural' environments, largely inhabited by endemic flora and fauna, but that is by no means the case.

Over the past 800 years the ecosystems that make up this landscape have changed greatly, and this process is continuing. Not only are the community boundaries shifting, but most of the grasslands to the east can hardly be called 'indigenous' or 'native', despite their appearance. The farmed valley floors are largely covered with exotic plants, and the vegetation on most of the hill country up to about 1,000 metres is a mix of native tussock and introduced grasses and herbs.

These changes are the result of the arrival of organisms from overseas, and by far the most important of these intruders are humans.

The first arrivals

Early Maori left few records of their occupation of the high-country valleys, and the evidence for changes during the 800 years of their occupation relies largely on the result of scientific research – the presence of charcoal in soils, pollen analyses of peat bogs or material washed out by rivers and deposited during those times, or analyses of the middens of moa-hunting Maori.

These indirect types of evidence suggest that when Maori arrived the vegetation pattern of the high country was closely related to the physical factors of the environment: rainfall, temperature and soil type. Beech was much more extensive on the hillsides – the forests now confined to the Torlesse and Craigieburn Ranges may have extended over most of the hills in the central Waimakariri region. A similar situation probably existed in the upper Rakaia. By contrast, the tussock grasslands were much less extensive, largely confined to river terraces and the floors of the basins and valleys in drier areas. The evidence shows that shortly after the arrival of Maori there was a significant increase in the frequency of fires, forests were reduced greatly and grasslands increased.

We can only speculate as to why the forests were burnt. It may have been part of the process involved in the widespread hunting that led to the extinction of the moa about 400 years ago. These huge ratites were almost certainly common in the high-country grasslands, as were weka, the other ground-dwelling birds that provided a food supply for early Maori.

By the time the first European pastoralists entered the high country, moa were long gone and tussock grassland had replaced large areas of forest in the east and central areas.

The pastoralists' passion for fire

The next phase of large-scale landscape change began in the 1860s, when Europeans settlers brought sheep to the high country. Fire was routinely and often indiscriminately employed to clear scrub and speargrass, and, more importantly, to induce succulent new growth in the tussocks. The subsequent grazing by sheep was another factor that the vegetation of the tussock grasslands had to cope with. In 1863 Samuel Butler recognised that 'the difference between country that has been fed upon by any livestock, even for a single year, and that which has never yet been stocked, is very noticable'.

He also noted that burning was considered essential to establish good pasture:

> The fire dries up the swamps – at least many disappear after country has been once or twice burnt; the water moves more freely, unimpeded by the tangled and decaying vegetation which accumulated round it during the lapse of centuries, and the sun gets freer access to the ground.

This view across the Rakaia Valley to Mt Hutt shows how extensively exotic plants have changed the landscape. Newcomers include shelterbelts, the cultivated terraces and, in the foreground, the remains of invading gorse. Indigenous ecosystems are largely confined to the slopes of the mountain.

'The Exceeding Joy of Burning'

In 1867 Lady Barker was living on a sheep run in the Malvern Hills. Her book *Station Life in New Zealand* includes a chapter titled 'The Exceeding Joy of Burning', which gives a graphic description how much she loved this activity and the indiscriminate way in which it was carried out, with no regard to long-term consequences.

> When [Frederick Broome, her husband] pronounces the wind to be just right, and proposes that we should go to some place where the grass is of two, or, still better, three years' growth, then I am indeed happy. I am obliged to be careful not to have on any inflammable petticoats, even if it is quite a warm day, as they are very dangerous; the wind will shift suddenly as I am in the very act of setting a tussock a-blaze and for half a second I find myself in the middle of the flames . . .

Later she describes sharing her expeditions with a like-minded pyromaniac:

> As soon as we come to a proper spot we begin to light our line of fire, setting one large tussock blazing, lighting our impromptu torches at it, and then starting from this 'head-centre', one to the right and the other to the left, dragging the blazing sticks along the grass. It is a very exciting amusement, I assure you, and the effect is beautiful, especially as it grows dusk and the fires are racing up the hills all around us. Every now and then they meet with a puff of wind, which will perhaps strike a great wall of fire rushing up-hill as straight as a line . . .

Burning and grazing soon began to alter the composition of the tussock grasslands, as explained in the previous chapter. Fire continued to be widely used as a regular farm-management practice through most of the first century of settlement. However, concern gradually began to grow about the possible effects of destroying useful shelter species, opening up areas to the entry of undesirable weed species and accelerating erosion. By the 1940s the use of fire was beginning to be seriously questioned, and its use has since declined greatly. Some farmers, however, still regard burning as a way of removing rank growth that is smothering pasture plants, though they are mindful that it needs to be carefully managed because of the danger it poses to neighbouring land and the fact that it can encourage the establishment of undesirable weeds.

The spread of exotic plants

Along with sheep and other introduced animals, a wide range of plants accompanied the European settlers to New Zealand. Some of these were deliberately sown in the high country, and others were accidental arrivals that quickly spread in their new home.

A close look at a tawny high-country hillside is likely to reveal that what appears to be a typical native grassland community is not quite what it seems. In fact, tussock may cover as little as 10 per cent of the surface, and most of the rest of the vegetation comprises exotic grasses. Early photos (see, for example, page 66) show that the effects of repeated burning and heavy grazing created much open ground in the tussocklands, having removed the protective mantle of decaying leaves and drastically reduced the palatable native grasses and herbs.

The exposed conditions favoured the entry of exotic grasses that are free-seeding and have a twitch-like rooting system. The two species that quickly became widespread were browntop (*Agrostis tenuis*) and sweet vernal (*Anthoxanthum odoratum*), and these remain dominant in the higher grasslands of the region.

Lower country, and especially areas that have been topdressed, have been colonised by less hardy but better-quality pasture grasses, including cocksfoot (*Dactylis glomerata*), Yorkshire fog (*Holcus lanatus*) and white clover (*Trifolium repens*).

Burning the tussocklands and creating more open space also encouraged invasion by a number of exotic weed species. Notable among these are sheep's sorrel (*Rumex strigosum*) and catsear (*Hypochaeris radicata*), but about 25 species of exotic herbs and grasses have now colonised the alpine tussock grasslands, and over 45 the short tussock areas.

In recent years the grazing viability of some parts of the South Island high country has been threatened by the spread of hawkweed (*Hieracium* spp.). Once established, this small and seemingly harmless plant forms tight, low mats that can exclude all other plant species, and it is extremely difficult to eradicate. The most common species are king devil (*Hieracium praealtum*) and mouse-ear hawkweed (*H. pilosella*). Although infestation in the Canterbury high country is not as serious as it is further south, patches in drier parts of the Lake Heron Basin are causing concern.

The exotic grasses browntop and sweet vernal now co-exist with and even replace the tussocks as the dominant plants in much of the high country.

One of the smothering hawkweed (*Hieracium*) species.

Invaders on show

The most obvious of the exotic plant invaders are several shrub species. Fifty years ago the spread of sweetbriar was recognised as a serious problem, together with Russell lupins in the riverbeds. Ironically, both had admirers who saw them as adding beauty to the landscape. Neither pest has been eliminated and they have been joined by another pair of colourful pests – broom and gorse – and by wilding conifers, especially pines.

The sweetbriar, or briar rose (*Rosa rubiginosa*), is particularly noticable in autumn, when its bright red rosehips put on an attractive display. Its seeds are readily spread by birds, and this plant thrives on drier grasslands, especially on shingle flats. Growing as a tangled mass, it forms an impenetrable barrier, often mingled with matagouri and other native shrubs. Sweetbriar is present in all of the Canterbury valleys, though it is considered a serious problem only in some localities, such as around the edge of Lake Pearson.

Sweetbriar rosehips.

Although attractive when in flower, Russell lupins (*Lupinus polyphyllus*) can be aggressive once they are established. They flourish on shingle riverbeds, thus reducing the feeding and open nesting areas available to wading birds and increasing their risk of predation. So far lupins remain a localised problem in the region.

Broom and gorse, originally introduced as hedging and ornamental plants, have colonised waste areas throughout New Zealand, and the Canterbury high country is no exception, especially where broom is concerned. In the hill country on the edge of the plains, both pests have spread to such an extent that, in some areas, they have excluded most other vegetation. Their presence frequently indicates the location of early settlements; for example, on the hills along the south bank of the Waimakariri Gorge, in the vicinity of the railway line, having spread from the sites of construction camps. On the lower slopes of Mt Somers, infestations are associated with the Blackburn coal mine and other early settlements.

Each of the gorges of the three large rivers has a good deal of dense gorse (*Ulex europaeus*) and broom (*Cytisus scorparius*) where it emerges from the hills,

Russell lupins at the western end of Lake Coleridge.

Gorse in flower in the upper Rakaia Valley. This heavy infestation has completely taken over the drier parts of the riverbed, excluding all other plants but a few wilding pines.

and both are common in the beds of the Kowai and Ashburton Rivers. Further inland, the only serious penetration is up the Rakaia Gorge, where dense thickets cover many of the more stabilised areas of the riverbed, particularly between the Lake Coleridge village and Mount Algidus Station.

The bright yellow flowers of these invaders makes for a spectacular sight, but the price of this is high. Some clumps of broom and gorse have become so large and dense that they are now a serious problem that will require a great deal of money and effort to eradicate.

Exotic trees

In many settled areas of the high country, and prominent among the tussock and other pasture, are dark lines of shelterbelts running across valleys and clusters of trees surrounding station homesteads. Many of the latter were planted quite soon after settlement and now form impressive stands of hundred-year-old specimens. Some of these, such as at Snowdon, contain a variety of big conifers, but the majority of exotic trees in the region are *Pinus radiata*, planted from the mid-1940s on as shelterbelts to reduce the ferocity of nor'westers at ground level.

An intriguing plantation of exotics can be seen in the village associated with the Lake Coleridge hydro-electric station, and on the hill behind the power-house. Planting began in 1911 shortly after work began on the project, at first following the initiative of the first engineer in charge, a Mr Kissell, and within 20 years over 150,000 seedlings (mostly *Pinus radiata*, *P. ponderosa* and Douglas fir) were growing. The main force behind the planting was Harry Hart, who was in charge of the power station for over 30 years. He

Interior of the Harry Hart Arboretum at the Lake Coleridge power station village. This collection includes nearly 150 different species of conifers, most about 70 years old.

Windbreaks of radiata pine across the line of the Rakaia nor'wester, one of which is brewing in the mountains beyond.

Experimental conifer planting

This 1953 photo shows a number of exotic conifers above the bushline on the high slopes of the mountain above Cora Lynn in the upper Waimakariri Valley. These were planted in 1947 by the North Canterbury Catchment Board in an experiment to determine how successfully various trees, including Douglas fir, larch and seven species of pine, would grow on sunny slopes at an altitude of about 800 metres. It was eventually concluded that conditions were too marginal for economic forestry.

is best remembered for the arboretum of coniferous trees (including two-thirds of the world's pine species) that he began planting about 1930. Unfortunately, the plantation has been neglected in recent years and the grounds are in poor condition.

A serious problem in most wilderness grasslands in New Zealand is the spread of wilding pines. Because of their abundant wind-blown seeds, most conifers can disperse over wide distances and thus they have become a cause for concern in the high country.

The spread of the wilding conifers from Castle Hill village is apparent in the photos on page 66. More difficult to eradicate are those around the fringes of the Craigieburn Forest at the northern end of the Castle Hill Basin. These are visible on the slopes facing the highway but are more obvious from the roads leading to the skifields.

Pines are seeding rapidly at the upper limits of the beech forest and are even spreading across scree slopes. They are much faster growing than beech trees and are rapidly taking over some of the subalpine ecosystems. As increasing amounts of mountain land are being passed to the Department of Conservation for long-term management, controlling the spread of wilding pines will be one of the department's most urgent tasks.

Willows add autumn colour along the banks of some streams and around the larger lakes such as Heron, Camp, Clearwater and Pearson, but the high-country climate appears to have prevented their becoming the problem they have become in some parts of New Zealand.

Other significant deciduous trees planted in the region, other than garden specimens around homesteads, include rows of Lombardy poplars, especially evident in the middle Rangitata Valley and at Flock Hill in the Waimakariri Valley.

This photo of beech forest growing on the lower slopes of the Craigieburn Range also shows pines in the middle ground, where they are spreading around the edge of the forest.

Willows on the shores of Lake Camp during winter.

The changing picture

In 1953, as part of a university research project, I attempted to gauge changes in high-country vegetation by comparing early photographs with others taken 70 years later.

1882: The view towards the Enys homestead in the Castle Hill Basin with the Craigieburn Range in background. The gate was designed for use in deep snow. Note the open spaces between tussock clumps, the result of fire and grazing.

1953: Exotic trees are growing around the old homestead site, much tussock has been replaced by browntop and sweet vernal grasses, and scrub (mostly matagouri) has spread. At the time it was thought that accelerated erosion was occurring on the hillsides, but there is no clear evidence of depletion of vegetation on the Craigieburn Range.

2005: Self-seeding larches put on a colourful display in autumn but have spread so much that it is impossible to take a photo from the same site. This view, from a hundred metres back, shows how the flatter land has been developed and is cultivated for winter feed.

A snow gate at Castle Hill Station, 1882.
Burton Brothers photograph, Canterbury Museum, 19XX.2.862

The same view in 1953 (***above***) and 2005 (***below***).

Introduced animals

From the time of their arrival the European settlers of New Zealand were keen to introduce animals from their homeland, partly for sentimental reasons but also for food and sport. In 1864 the Canterbury Acclimatisation Society was set up for this purpose, and the first introduced mammals and birds arrived in the high country not long after the first runholders. As far as records show, the only non-domestic animals deliberately released in the region were red deer and several species of game fish.

Although nearly 20 species of mammal and a similar number of exotic bird species reached the high country, only a few have made a lasting impact and the area has been less affected by introduced animal pests than highlands to the north and south.

In spite of being introduced to Canterbury as early as 1866, rabbits never reached the plague numbers in the valleys of central Canterbury that they achieved in areas such as Molesworth and the Mackenzie Basin in the south. In some drier districts, however, they were regarded as a serious problem. In the early 1920s rabbits became such a problem on Mesopotamia that over 150,000 were taken off the property in one year. They were eventually controlled by the use of poisoned carrots. In the upper Rakaia a rabbit board was set up to deal with these pests, which caused much concern in the Lake Heron Basin. On Clent Hills in the mid-1920s the new owner, Bob Buick, spent two years eradicating rabbits before he was able to put sheep onto the property. Rabbits multiplied hugely in the area during the Second World War, and in the late 1940s up to 20 rabbiters were employed on Mount Possession. It was only when that run's innovative runholder Sam Chaffey spread poison bait by aircraft that the rabbits were finally controlled. Since that time intensive pest control and pasture improvement by topdressing have kept the population down.

Hares were reported as escaping from a ship in Lyttelton Harbour in 1851. They gradually spread to the high country, where they adjusted to somewhat higher altitudes than rabbits. By the 1940s they were reported in light to moderate numbers, especially in the Rakaia, Rangitata and Ashburton Valleys, and are still seen occasionally.

Small numbers of pigs have been present in parts of the eastern hill country for many years – Lady Barker described a pig hunt in the 1860s – but they have never been widespread.

Twenty-one Australian brush-tailed possums were liberated by the Canterbury Acclimatisation Society in 1865 in the hope that they would be of economic value for their fur. Although relatively common in high-country forests and around homesteads by the 1940s, they were not regarded as a serious pest. More recently, possums have become a major problem in the drier beech forests, where they do a great deal of damage by feeding on the more palatable plant species and raiding birds' nests.

Game animals

The most significant of the game animals released in the high country were red deer. Nine animals from an estate in Buckinghamshire, England, were released as early as 1897 in the Rakaia Valley and a second herd was established in the Poulter Valley on Mount

Pesky possums

For several years in the early 1940s my brothers and I were fortunate to spend part of our summers camping in the bush at the base of the Big Ben Range with our parents. We slept in a tent under beech trees, on mattresses of 'mikimiki' – springy coprosma branches. These holidays must have been hard work for my mother, but we boys enjoyed an idyllic time. One of my vivid memories, though, was as a very scared eight-year-old being woken in the middle of the night by possums screeching loudly from a branch just above the tent, then dropping onto the fly and sliding down to the ground.

Camping on the edge of beech forest in the 1940s.

Deer being transported in carts to be released in the Poulter Valley in 1909.

P. Sparkes photograph, Canterbury Times Bishop Collection, Canterbury Museum 1923.53.435

White Station in 1908–09. Both herds thrived and in the 1920s were thought to comprise several hundred animals, though they were widely separated from other South Island herds, in Nelson and at Hawea. A 1947 survey reported light populations of deer (one to two seen at a time) right through the forests south of the Rakaia, but much heavier infestations (10 or more seen at a time) further north.

Red deer need cover and generally live around the beech forests, though they have adapted well to feeding above the bushline and in scrub and tussock. As their numbers increased, deer caused increasing damage, especially in the drier beech forests, where they severely depleted understorey plants. By the late 1930s deer had earned the reputation of being a major pest, particularly in the headwaters of the Rakaia and Rangitata, and government-organised culling operations were set up. One of these resulted in three cullers shooting over 3,000 deer and 1,000 chamois in three months at Erewhon Station.

A large bull tahr.

Courtesy Back Country New Zealand

This scale of eradication by the Deer Control Section of Internal Affairs, then the Forest Service from 1956, soon depleted the wild herds. More recently, control has been maintained by recreational hunters. Ironically, the red deer population has burgeoned again on some properties, but the newcomers pose little threat as they are farmed behind high fences.

Tahr and chamois were released early in the twentieth century in the Mt Cook area to provide sport. Tahr, originally from the Himalayas, live in high rocky bluffs and have spread to the mountains south of Arthur's Pass. When present in large numbers they are a threat to fragile alpine flora, and the Department of Conservation periodically conducts culls to keep numbers below 10,000. Chamois, from the European Alps, thrived above the bushline in their southern counterparts and by 1947 had spread through the central Canterbury high country, especially in the Rakaia headwaters. Both animals are highly prized by recreational shooters and attract many trophy hunters from overseas.

Other exotic animals

The high country has been invaded by several small predatory mammals such as ferrets and stoats, and also feral cats. These are not often seen, but they pose a considerable threat to bird life.

A number of introduced insects and other invertebrates are present in the regions. They have had little impact on existing ecosystems, with one notable exception – the European wasps that invaded the drier beech forests after the Second War War. Their effects are discussed in some detail in Chapter 3.

It is easy to overlook the fact that most of the birds that are so much a part of the life around the homesteads and grasslands of the high country have not always been part of this environment. The Canterbury Acclimatisation Society was certainly successful in its aim of surrounding the early settlers with the familiar birds of England. Before the end of the 1860s at least 15 species had been introduced.

Many of these quickly reached the high country, and sparrows, thrushes, blackbirds and starlings became common around the homesteads. A variety of finches thrive on the grasslands and the endless wavering song of the skylark is heard throughout the summer. Two Australian birds also thrive in the lower country: the magpie was introduced to keep insects in check; the spur-winged plover crossed the Tasman unaided. The Canada goose is the only exotic bird to have become a pest: it causes damage to pastures and winter feed in wetter areas.

Although early attempts to import live ova from the northern hemisphere freshwater game fish were fraught with problems, later introductions of brown trout, rainbow trout and quinnat salmon quickly became established and displaced some of the native galaxids and other small fish. The story of salmonids in the high country is covered in Chapter 8.

Poplars at Flock Hill in autumn.

Nature in change in the high country

Since humans arrived in the central Canterbury high country the landscape has undergone changes – some dramatic and wide-ranging; others more subtle and localised. These began with the arrival of Maori about 800 years ago, and intensified and accelerated when Europeans explored then settled the area from the 1850s. The valleys and mountains now contain a blend of indigenous communities and a wide variety of exotic plants and animals, including humans, that have modified the region.

Although various exotic flora and fauna have brought about great changes in the high country, it is worth remembering that by far the greatest impacts on the land are the direct result of human invasion. Our signs are everywhere!

CHAPTER FIVE

Early Settlement

Human invaders

The story of human involvement in the catchments of the three great Canterbury rivers begins about 800 years ago when Maori food-gathering and hunting parties made expeditions to the area and travellers passed through it on their way to the West Coast. European explorers first penetrated the interior in the late 1840s, and by the end of the next decade leases on most of the high country had been taken up by ambitious pastoralists.

The Maori presence

All of the nine traditional Ngai Tahu settlements in what is now Canterbury were located near the coast, close to food resources, and there is little evidence that Maori permanently inhabited the high country. However, they did have two good reasons for short-term forays into the area.

Seasonal food-gathering expeditions extended as far as the inland lakes to obtain eels, weka and water birds. It is likely that moa-hunting expeditions also ventured into these basins. Evidence from charcoal in the soil shows that at this time there was considerable destruction by fire of forest and scrub, which was formerly much more extensive in the high country. It is thought that some of these fires were lit by Maori, perhaps in association with hunting.

Maori food-gathering visits were made during the summer and tended to be of quite short duration. Cave paintings on the limestone rocks at Castle Hill, dated at about 500 years old, indicate that this was an area of spiritual significance, and the discovery of a flax backpack suggests it was a stopping place for travellers. Elsewhere, the evidence of Maori presence is not abundant: cooking ovens have been unearthed at Bealey, camping places around Lake Coleridge and an adze near Double Hill in the upper Rakaia.

The second reason for Maori visiting the high country was to obtain the much-prized pounamu, or greenstone, from its source in the rivers of the West Coast. Regular journeys from coastal settlements such as Kaiapoi were undertaken over the Southern Alps on this quest. The usual route taken was from Lake Sumner over what is now Harper Pass and down the Taramakau River. Several other trails further north and south were also used. These were relatively easy treks for the parties of up to dozen people, including women (who were the carriers), children and slaves. The much more difficult Browning Pass, at the head of the Wilberforce River, and Arthur's Pass were attempted by raiding parties of fit males who were looking for speed and surprise. Even the very high Mathias, Whitcombe and Sealey Passes are known to have been used, despite the likelihood of encountering extreme weather conditions. Maori tales refer to

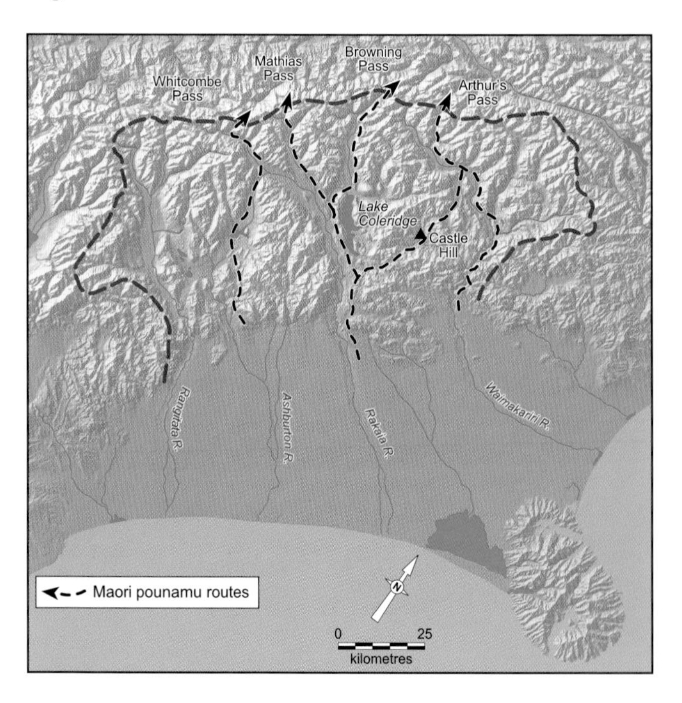

Routes taken by Maori on journeys to and from the West Coast.
Cartography by Tim Nolan

The spectacular limestone rocks at Kura Tawhiti (Castle Hill), which has long been regarded by Maori as a sacred site.

frequent hardship and exposure resulting from storms during these difficult alpine crossings.

Lake Coleridge, known to Maori as Whakamatau, was certainly a frequently used stopping and food-gathering place for expeditions to the West Coast. Its significance was recognised by its inclusion in the Ngai Tahu land claims settlement in 1998.

These early Maori travellers appear to have been well equipped for mountain treks, and remains of plaited flax backpacks and rain capes have been found. There is no doubt that the high-country valleys were familiar to Maori, as they had named many features in the area. It is interesting to note that the first European settlers named the three great Canterbury rivers the Courtenay, Cholmondley and Alford, but within quite a short time they reverted to their traditional Maori names – Waimakariri, Rakaia and Rangitata. European explorers were impressed by the ability of local Maori to recite place names and to draw rough maps of routes, and they were regarded as reliable guides.

Maori mythology

Maori mythology and legend directly related to the central Canterbury high country is sparse. One tradition tells of the chief Rakaihautu, who captained the waka (canoe) that brought the Waitaha iwi (tribe) to Aotearoa. They landed in what is now Nelson and then travelled through the interior of the South Island. Along the way Rakaihautu used his ko (digging tool) to carve out a string of high-country lakes, including Whakamatau.

Another Waitaha story features a great bird, Pouaki, that had a nest high on Mt Torlesse. It terrorised the people of the plains by descending at great speed to carry off women and children as food for its young. A warrior named Te Hou o Tawera offered to get rid of the monster. One night he and 50 men covered a great hole with interlaced manuka branches, under which the warriors hid. In the morning Te Hou o Tawera emerged to lure the bird from its nest. As Pouaki swooped on him, he leapt into the hole, the bird followed, became entangled and, after a great fight, was killed by the spears of all 50 warriors.

Tradition has it that Waitaha were eventually exterminated by Ngati Mamoe, who in turn were conquered by Ngai Tahu. The newcomers' discovery of pounamu is also the subject of a legend. While wandering in the mountains, Raureka, a woman from Ngati Wairangi on the West Coast, is said to have found and crossed what is now Browning Pass into the Wilberforce Valley. She showed a piece of greenstone to some Ngai Tahu who were working on a canoe with adzes. They were so impressed by this stone, which could be honed to a razor-sharp edge, that a party was sent across the mountains to investigate its source. In the wars that followed, Ngai Tahu gained control of the pounamu areas of the West Coast and were in a position to trade this valuable commodity with North Island iwi.

The scramble for pastoral leases

The forbidding mountains and gorges rising west of the Canterbury Plains held little attraction for the European pioneers who arrived in the province in the 1830s. Even five years after the Canterbury Association settlement in 1850 there had been just a single attempt to explore the high country. On New Year's Day 1849 the surveyor Charles Torlesse, accompanied by a young Maori, George Tuwhia, climbed Otarama, one of the summit peaks in the range that would later be named after him. From there Torlesse saw Lake Coleridge and the snowy peaks up the Rakaia Valley but, with few exceptions, that vista failed to capture the imagination of Europeans for another six years.

A brief diversion is needed to explain the situation regarding farming land in the province before interest in the high country was awakened. By 1850 the Canterbury Association had purchased from Maori the 'Canterbury Block', which comprised most of the plains and all the high country to the west. The association's original vision was to recreate the English agricultural farming system, based on well-to-do owners of freehold property employing waged labourers. But the combination of expensive land, small markets and labour shortages meant that this form of farming was slow to develop in Canterbury. At the same time, the Australian experience had shown that 'squatting' – pastoral farming with large flocks of sheep on large blocks of land available at nominal rentals – had proven profitability. Canterbury's land-tenure system was thus modified, providing for leases that could be taken up very cheaply.

This resulted in a scramble for blocks of land, and there are many stories of races to register boundary points and apply for grazing licences. Generally, the lessee had no preemptive rights to the land, and at any time someone else could buy up a portion and freehold it. Indeed, some people did purchase small patches of the best land, ensuring that the surrounding area was of little use to another potential leaseholder. The practice of 'gridironing' – whereby a series of blocks scattered over a property were freeholded – also caused problems.

In spite of such frustrations, Canterbury's new settlers were eager to begin farming, and by 1854 grazing rights had been taken up for almost all of the plains and most of the land there was freeholded. Graziers were also moving into the Malvern Hills, between the Rakaia and Waimakariri, and taking up runs such as Dalethorp, Rockwood and Steventon (later to be

William Rolleston's 1861 homestead at Rakaia Forks is typical of the buildings erected by early runholders. It was constructed of totara slabs with a thatched tussock roof. Although remote, this was one of the earliest properties established in the Canterbury high country. Between 1857 and 1860 four blocks of land were taken up between the junction of the Rakaia and Wilberforce Rivers, and in 1861 these were amalgamated by Rolleston. He ran sheep on the property for four years before selling it to Frank Neave, who renamed the station Mount Algidus and farmed it for the next 19 years.

Courtesy Corinne Crawley

William Rolleston, like many of the first runholders, was an English gentleman, cultured and well educated. He served as Canterbury's Provincial Superintendent 1868–76.

D. L. Mundy photograph, Canterbury Museum 19XX.2.860

made famous through the writings of Lady Barker.)

The snow-capped ranges that dominated the western horizons generally remained an intimidating barrier to those with farming ambitions. In 1851, however, Mark Stoddart and two companions ventured up the terraces on the north side of the Rakaia Valley to find the lake that Torlesse had glimpsed two years earlier. Having reached Coleridge, they turned back, setting fire to the tussock grasslands on the way. Although three sheep runs – High Peak, Snowdon and Acheron Bank – were soon established in this reasonably accessible area, it was widely believed that any land higher than the rolling downs could not be profitably stocked.

Explorers move into the high country

The splendid isolation of the mountains changed abruptly in 1855 when Charles Tripp and John Acland set out on a series of expeditions into the high country. The promising result of their endeavours began a frantic period of exploration, and within three years leases had been taken up on nearly all of the land in the three major river valleys.

In September 1855 Tripp and Acland set off up the south side of the Rangitata, travelling more than 30 kilometres over the previously unexplored tussock-covered hills as far as Forest Creek. Six months later they passed through the Ashburton Gorge and set out across the extensive flats on the north side of the Rangitata as far as Potts Creek.

Others were now inspired to look for land in the mountains. In 1857 Thomas Potts, his brother-in-law Henry Phillips and Francis Leach (owner of Snowdon Station) ventured further up the Rakaia, discovering a route to Lake Heron, and came out through the Ashburton River. Following these explorations, Tripp, Acland, Potts and Leach secured runs in this area, between them acquiring most of the land between Lake Heron and the south side of the Rangitata. Three years later, Samuel Butler began his explorations of the upper valleys of the Rangitata and established the run he named Mesopotamia.

Also in 1857, Joseph Hawdon, a South Australian pastoralist who was planning to migrate to New Zealand, employed Joseph Pearson to look for grazing land in the interior. An enterprising Cumberland man, Pearson loaded three packhorses with supplies and set off on a very difficult route around the northern end of the Torlesse Range, above the Waimakariri Gorge. (He was accompanied by a station hand, John Sidebottom, who had apprehended the sheep-stealer

Charles George Tripp

Charles Tripp arrived in Canterbury in January 1855 and worked as a cadet on Halswell Station. Later that year, in partnership with his friend J. B. A. Acland, he took up the lease on the vast area that would become Mount Peel Station. Subsequent explorations saw this pair obtain the leases to the Mount Possession and Mount Somers runs.

Tripp had a shrewd eye for country and showed great judgement in buying the freehold areas of his run at Orari Gorge.

John Acland.
F. E. McGregor photograph, Christchurch Club Collection, Canterbury Museum 19XX.2.857

Charles Tripp.
Canterbury Museum 19XX.2.858

John Barton Arundel Acland

John Acland accompanied Charles Tripp from England's West Country and joined him in exploring and laying claim to large tracts of land in the Rangitata and Ashburton Gorge areas.

In 1862 the partnership was dissolved and Acland became sole owner of Mount Peel, which has remained in family hands to the present day. At its greatest extent the run encompassed almost 42,500 hectares but is now much smaller since much of the land further up the Rangitata was taken by the government around 1912 for closer settlement.

Acland was a public-minded man, active in the local community and in national politics as a member of the Legislative Council from 1863 for more than 30 years.

James McKenzie at Burkes Pass the year before.) The pair emerged into the Broken River Valley and spent several weeks exploring and burning the tussock basins around Cass, Flock Hill and the upper Waimakariri. Guided by Pearson's advice, Hawdon took up leases over much of this country for runs that became Grassmere, Craigieburn and Riversdale.

In 1858 Charles Torlesse crossed over what is now Porters Pass to explore and map the Castle Hill Basin and nearby areas.

Following these expeditions there was a rush to file claims for leasehold properties, and by 1859 all but a couple of the most remote areas in the headwaters of the three major valleys had been taken up as some 30 large pastoral holdings. There were no more than 20 or so men involved in the initial applications, although in the next few years many of the leases were transferred to others.

The exploration of each large valley was carried out in a similar fashion and was well documented in the diaries of men such as Potts and Acland. Parties set off on slow and arduous horseback journeys through valleys covered with dense thickets of tall tussocks, flax and other vegetation, including an abundance of spiny spaniards and gnarled, thorny matagouri – both damaging to horses' legs.

Not only was the scrub hard to penetrate, but in that state the land was quite useless for sheep grazing, so these travellers set fires as they went up the valleys. L. G. D. Acland, in *Early Canterbury Runs*, recorded that in 1857

> [Joseph Hawdon] sent Pearson up the Waimakariri to explore. According to the custom of those days, Pearson burnt the country as he went, and though he was away longer than they expected Hawdon could see from the plains the smoke from his fires in the upper Waimakariri and knew that he was at work . . .

In fact, the glow in the sky beyond Mt Torlesse was a major talking point among Christchurch citizens and the fires lit by Tripp and Acland further south were reported to be visible 120 kilometres away.

Our Camp Near the Tasman Glacier, a watercolour by William Green. Although this scene is of the mountains south of the focus of this book, it shows the extent and indiscriminate nature of the fires lit to 'clean up' the hills.

Alexander Turnbull Library A-263-011

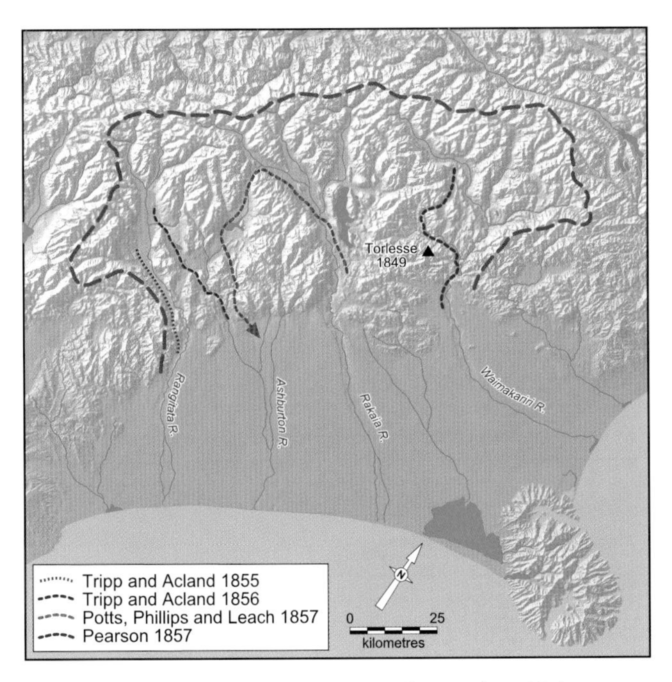

Routes of European explorers in the central Canterbury high country.

These conflagrations had an immense impact on ecosystems that in many cases had been undisturbed for perhaps hundreds of years. They left behind desolate, blackened hillsides, but when the rains came a mat of palatable grasses and herbs would spring up on the burnt areas. It was to be many years before the long-term adverse effects of soil erosion and invasion by weeds were recognised.

As the explorers worked their way through the high country, camping in the tussocks, they encountered hazardous river crossings and the ferocity of nor'west storms that regularly funnelled down the valleys. The diary of Thomas Potts records a dramatic incident when his party reached Lake Heron and set up camp:

> Harry [Phillips] went duck shooting, Leach and T.H.P. [Potts] cooking. Leach gave an alarm and T.H.P. ran to see what was the matter. The grass was on fire. We endeavoured to put it out, but in vain – the wind was too strong. We untied our horses, threw our swags on them, and ran into the riverbed waving everything but a waistcoat. We camped in the riverbed but could get no wood to spread our blankets with. The weather looked bad – very bad, but we were so pleased at finding this plain and a new route home by the Ashburton that we were happy enough.

During this journey the intrepid trio travelled up the Rakaia Valley, past Lake Coleridge to the Wilberforce River, where they camped the night on an island, sheltered beneath clumps of toetoe, after catching and cooking some ducks.

Over the following days they proceeded further up the Rakaia River and Lake Stream, to discover Lake Heron and explore the country to the south. There they found Lake Clearwater before venturing down the Ashburton River to emerge on the plains near the present Mount Somers village. Leach immediately applied to lease land in the Lake Heron Basin, and Potts staked his claim to a block further south.

Pilgrims and Prophets

Christchurch society recognised two types of settler who took up land in the early years of the province. A third group, comprising mostly Scottish shepherds, were largely unsung.

The 'Pilgrims' were the original settlers who had arrived on the 'First Four Ships' with the Canterbury Association. Men from the upper classes of English society, well educated and with money, they were expected to establish farms and form the backbone of the new colony. Many of them, however, were completely inexperienced, particularly at pastoral farming, and although some gleaned advice from the very early settlers on the plains, such as the Deans family, others learned the hard way or were reliant upon the shepherds they employed.

Some of the upper-class arrivals came to the colony to make money as quickly as possible and then move on. Among these was Samuel Butler, who arrived in New Zealand in 1860 with the aim of doubling his capital of £4,000. He had no experience of sheep farming but did have a good deal of common sense. In 1860 Butler explored the upper Rangitata and leased the block of land he called Mesopotamia. Three years later he sold his stock, achieving his desired profit, and returned to England. His book *A First Year in the Canterbury Settlement* is a fascinating record of the trials and tribulations of establishing a sheep run in the remote high country.

The second group who took up land were dubbed 'Prophets'. These were Australian squatters, often also referred to as 'shagroons', who prophesied the doom of the Pilgrims and their plans to develop small holdings of agricultural land. Although they had considerable experience of sheep farming in Australia, they generally did not have the ready money to buy large amounts of land. Many had suffered from severe droughts in Australia during the 1850s, and some crossed the Tasman with their flocks of sheep.

Samuel Butler

Butler was an English gentleman who arrived in New Zealand in 1860 with the aim of obtaining a run and quickly doubling his money by sheep grazing. When he found that there was no land left to lease on the Canterbury Plains, he introduced stock on the Mesopotamia run in the upper Rangitata Valley. The ferocity of the scramble to obtain leases is illustrated by his feud with J. H. Caton, a neighbouring runholder who had erected a hut on Butler's land. Both wanted to freehold the land and, in an attempt to be first to register a claim, they raced each other the 160 kilometres to Christchurch on horseback. When the Land Office opened, Butler found that Caton's solicitor had written his client's name at the top of the list. The dispute went to the Land Court, which ruled in Butler's favour.

In the three years he was in New Zealand, Butler not only achieved his aim of doubling his money, but also spent a great deal of time exploring the headwaters of the Rangitata as well as the Waimakariri, where he was the first European to see the alpine saddle that became known as Arthur's Pass and, with J. H. Baker, was the first to cross Whitcombe Pass. His 1872 novel *Erewhon* is set in a mythical land over these mountains.

Christchurch Club Collection, Canterbury Museum, 19XX.2.859

Joseph Hawdon

Hawdon was one of the first squatters to arrive from Australia, and perhaps the best example of a 'Prophet'. He came from Melbourne, bringing sheep with him, and his favourable reports of Canterbury triggered an influx of other drought-afflicted Australians. It was partly on his insistence that John Robert Godley, in his role as chief agent of the Canterbury Association, managed to alter the land-holding regulations to favour large leases.

Hawdon acquired property on the Canterbury Plains and brought in more stock from Australia. Then, in 1857, following Tripp and Acland's example in taking up a run in the interior, he sent his manager, Joseph Pearson, to explore the back country of the Waimakariri. On Pearson's recommendation Hawdon applied for leases to what became the Craigieburn, Grassmere and Riversdale Stations. From this huge block of land he made a fortune in the 1860s and later became a member of the Legislative Council.

Canterbury Museum 19XX.2.861

Life in this challenging country quickly broke down class barriers, and the successful runholders soon abandoned the customary distinctions between English gentlemen farmers and their employees. The shepherds, mostly Scottish, brought much skill and experience in managing stock and often later became run managers and runholders themselves. Sober men of strong character, thrifty, living solitary lives and reading extensively, they were a complete contrast to the other run employees – the bullock drivers, cooks and shearers – who frequently blew all of their wages on a binge in the city.

Setting up a sheep run

Setting up a farm in the midst of the mountains was a daunting task, but most of the early runholders were redoubtable pioneers. Initially, the would-be squatter had to come to terms with the environment. The exposed, open valleys and wide, barren riverbeds, surrounded on all sides by forbidding mountains, must have seemed quite threatening to people used to the closely settled rural landscapes of Britain. The tussock grasses were not particularly nutritious, and although there were palatable herbs and grasses growing among them, it was necessary to burn off the abundant speargrass and matagouri before the grasslands could produce suitable grazing for sheep. An extreme climate had to be withstood, with frequent nor'west gales and, in winter, heavy snowfalls that could cause widespread destruction to flocks of sheep. Isolation was a real problem in the absence of roads, and increased by the danger of crossing the larger rivers. Indeed, drowning was known as 'the New Zealand death'.

Acquiring stock was not easy. With many runs being opened up, sheep were in short supply and expensive to buy. While accompanying flocks of up to 1,500 ewes across the plains, drovers had to avoid these mixing with sheep from other stations on the route. Care had to be taken that the animals did not eat tutu, a common and poisonous plant of riverbeds. To prevent this threat, sheep were usually well fed before being driven quickly through known patches of 'toot'.

The runholder would generally travel with a bullock dray loaded with stores and tools, plus a number of horses and as many workmen as he could afford – a shepherd, a bullock driver and several men to help with fencing and timber cutting, and sometimes a cook. In *Sheep and Sheepmen of Canterbury*, Sheila Crawford describes the immediate tasks on arrival:

> Having reached the run at last, the squatter would first choose the site of his home station, bearing in mind such things as 'nearness to water and wood, accessibility, proximity to the place where the wool

'On the Road', a sketch by L. J. Kennaway, from his book *Crusts*. 'The sheep were so wild and restless that it was necessary to keep a double watch. The tent is a structure not a whit less rude than that shown in the sketch.'

Christchurch City Libraries

Above: The cob-and-thatch cottage built by Samuel Butler on Mesopotamia in 1861. At this stage there was only one dwelling and some yards set among the tussocks. Compare this and the painting below with the 1871 photo on page 113.

Alexander Turnbull Library F-386-1/4-MNZ

Below: A watercolour by William Packe showing buildings at Mesopotamia in 1868.

Alexander Turnbull Library A-196-015

shed must be, a central position, capability of shepherding from it two flocks, good garden soil, natural facilities for making paddocks'. Then after the tents were pitched, or grass whares, 'not much bigger than dog kennels', were built, sheep yards with post and rail or sod fences would be made. Next a stronger dwelling place would be built and possibly a wool shed.

Temporary yards were often built of scrub to hold the sheep at night until they were used to the area. Shepherds were required to be constantly vigilant because, apart from some natural boundaries such as major rivers or steep ridges, there were no fences on the runs.

Better houses would gradually be built – perhaps of cob with a roof of thatch, or of totara slabs if there was bush nearby. A garden would be established, paddocks fenced for horses and rams, or for growing grain, and a sheep dip to combat scab. Several early photos and watercolours done by Victorian amateur artists illustrate these early homestead sites.

Life on the early high-country runs was hard and the work was often monotonous, as was the food.

An early high-country homestead

The Castle Hill run was taken up in 1858 by the Porter brothers and sold six years later to John and Charles Enys and Edward Curry. The development of the property over the next 26 years was recorded in the diaries of the Enys brothers and a series of watercolours by Charles. The original Enys homestead, Trelissick, was sited near the present Castle Hill village.

Trelissick homestead, built in the 1860s at Castle Hill.

The interior of Trelissick sketched by William Packe in 1868. Contrasting with the slab timber construction and the simple furnishings is a large library.

Alexander Turnbull Library NON-ATL-P-0089

Sketch by Robert B. Booth in his *Five Years in New Zealand* of snow-raking for buried sheep.

Christchurch City Libraries

Such activities as fencing (at first putting up post-and-rail yards around the homestead, then wire-fencing paddocks as they were developed), lambing, burning off the tussock, snow-raking in winter, dipping sheep and carting wool out kept the station hands busy. Mustering and, particularly, shearing were seasonal activities that involved obtaining a large gang of additional men.

Apart from the labour and hardship involved, the costs of starting a station were considerable. In 1858 John Acland reported on a visit to England that at least £2,000 would be needed:

> . . . A thousand ewes would cost £1,000, a dray and eight bullocks, tools, household utensils, first year's stores and the expense of putting up a house, etc., would cost over £500 and it was necessary to have £500 pounds in reserve to meet wages and expenses for the first one or two years.

For the first few years of settlement the high country was almost entirely a male domain – there are virtually no records of women in the area before about 1860. When runholders did begin to bring wives onto the stations, this was the catalyst for the building of a reasonably substantial homestead and improvements to its surroundings. Even so, these pioneer women frequently endured a primitive way of life, having to keep house in formidable conditions and often suffering great loneliness.

Lady Barker provides a most entertaining account

This lonely hut was the home of George McRae and his wife during the 1880s when they had the property then known as Stronechrubie in the upper Rangitata Valley.

Courtesy Colin Drummond, Erewhon Station

of life on an early sheep run in the books she compiled from her diaries and the letters written to her sister in England. Although Steventon, in the Malvern Hills, would now hardly be classed as high country, it experienced most of the problems typical of runs further in the interior. The property had been taken up by 1854, but it was not until 1866 that her husband, Frederick Broome, and Henry Hill took it over. Lady Barker's arrival on the property followed the building of a homestead, Broomielaw, and prompted the extensive planting of trees and gardens and the arrival of young women as servants.

A much more difficult life must have been the lot of the less well-connected women who lived on the remoter runs. Perhaps the most poignant indication of the hardships endured by some of these pioneering women is the story of the wife of George McRae of Stronechrubie. This run was located at the junction of the Clyde and Lawrence Rivers, about five kilometres upriver from the present Erewhon homestead, and nearly 60 kilometres by rough track from the nearest village of Mount Somers. McRae was forced to work elsewhere much of this time, and it is said that his wife did not see another woman for 10 years. Eventually she had a mental breakdown and was taken from Stronechrubie, never to return.

Approximate location of the initial leases in the high country of central Canterbury. This is based on L. G. D. Acland's *The Early Canterbury Runs* and shows about 35 runs (depending upon where the boundary between high country and plains is drawn) occupying almost all of the area except the mountains near the Main Divide. The pattern has changed greatly, as Chapter 7 shows.

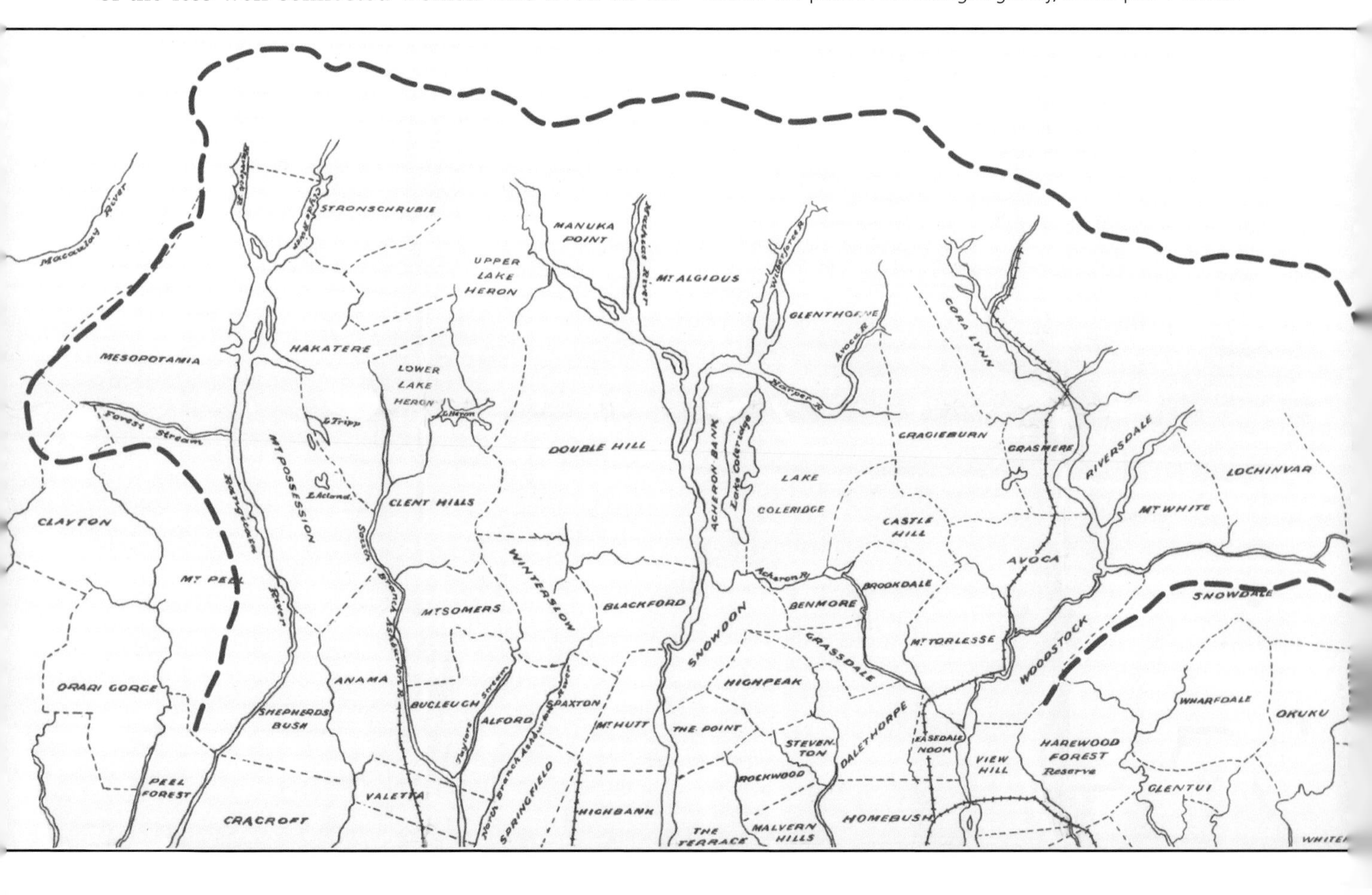

An 1870s watercolour by C. D. Barraud showing a sheep-washing race on Mount Alford Station. This task, commonly carried out before shearing, significantly reduced the weight of wool to be carted long distances by bullock wagon.
Courtesy Bruce Tinnock

Success and failure

Some of the early runs were immediately profitable, but by the 1860s the squatters' fortunes were under threat. Stock loss from tutu poisoning was always a danger on some properties, and a series of heavy snowfalls had disastrous consequences. Steventon Station lost 4,000 of its flock of 7,000 sheep in the storm of 1867. Another major problem was scabies, a highly contagious disease causing severe irritation to the skin of sheep. In 1857 several mobs of scabby sheep were brought down from Nelson and infected many Canterbury flocks. Fines were imposed on owners whose sheep passed on the disease, but it continued to spread rapidly and caused the ruin of several runholders. Regular dipping was found to be the best preventative, and this became a major undertaking before the advent of fencing to isolate the runs.

Some of the lessees found pastoral farming too costly an enterprise, runs changed hands quite frequently and, by the 1870s, many had been taken over by loan companies. Although Mount Peel Station has remained in the Acland family for almost 150 years, the ownership pattern at Mount White, described on page 82, is more typical.

Eventually, however, most of the high-country sheep stations became established so that by the end of the nineteenth century all the land that could be used for grazing sheep had been taken up, mostly as pastoral leases. These had become reasonably fixed, the environmental conditions better understood and appropriate farming methods had become stabilised.

Later explorations

It was not until some years after most of the high country had been taken up by squatters that explorers ventured as far as the main range of the Southern Alps. In 1863 the Christchurch surveyor John Whitcombe crossed the pass, later named after him, with a Swiss mountaineer, Joseph Lauper. They were poorly

Mount White Station

The early history of the occupation of Mount White illustrates the ups and downs of many of the sheep runs as they were becoming established. The remote hill country round the Esk and Poulter Rivers on the north side of the Waimakariri was first taken up in 1857–58 by E. C. Minchin as four separate leases with a total area of about 12,000 hectares. An Irishman who had arrived in Canterbury in 1853, Minchin continued to live in Christchurch (a not-uncommon practice among leaseholders) and left the management of the run to his two sons. He made frequent trips back to Britain and eventually returned there permanently in 1880.

In 1860 the station was sold to Major Thomas White, whose brothers ran the property, now named after the family. A new house was built at Lake Letitia, where the present homestead is situated. White's venture failed in 1864 and Minchin briefly took it back before it was sold in 1870 to J. M. Cochran. Again Mount White proved too big a burden to run, and Cochran committed suicide in 1884. The property was taken over by the NZ Loan and Mercantile Company, which held it for 18 years. Horses bred on the run were used for hauling trams in Christchurch and they continued to be noted throughout much of the last century as pack horses and farm hacks. Appropriately, Mount White is now one of the stopovers for horse treks that travel from North Canterbury to Tekapo.

The station changed hands again in 1902 and was finally given a long period of stability when, in the 1920s, it was bought by the Turnbull family, who have continued to own it to the present day. During this time the property has been run by a series of managers and has built up a reputation as one of the iconic high-country sheep runs.

This watercolour of Lake Letitia and the Mount White woolshed was painted in the 1870s by Charles Enys, of Castle Hill Station. (See also photo on page 104.)

Canterbury Museum 1957.26.3

prepared and equipped, and they had great difficulty reaching the West Coast. Whitcombe eventually was drowned while trying to cross the Taramakau River.

Exploration of the headwaters of the Canterbury rivers was activated by the discovery of gold on the West Coast in 1865. Immediately a search was undertaken for a suitable route across the main Divide, and several surveyors explored up the headwaters of the Wilberforce and Waimakariri Rivers. They found the Harman and Browning Passes, both of which seemed too difficult, but in the same year Arthur Dobson negotiated the crossing that would be named after him (see page 85).

Before the West Coast road was put through, the route generally followed by miners heading for the diggings went up the north side of the Rakaia River, climbed over Coleridge Pass into the Porter Valley to Castle Hill, and followed the line of the present main road up the Waimakariri and Bealey Rivers, over Arthur's Pass to the West Coast.

In 1882 traces of gold were discovered near the top of Browning Pass, and for the next three years a small group of men lived in this extremely inhospitable area, attempting to find the gold seam. There can have been few more miserable occupations than this in the early history of the central Canterbury high country.

P. Shourup photograph, Canterbury Museum 19XX.2.736

Julius Haast

The story of exploration of this region must include Julius Haast, a self-taught geologist from Germany who came to New Zealand in 1858 and was appointed provincial geologist for Canterbury in 1861. He made a series of expeditions into the Canterbury high-country valleys, exploring and investigating their geology: the upper Rangitata in 1861, the Lake Heron area in 1864, the upper Waimakariri in 1865 and the Rakaia Valley in 1865–67. Haast made the first accurate maps, drawings and geological studies of the region, including recognising the role that ice played in shaping the land, and was responsible for the names of many landscape features in Nelson, Canterbury and Westland. He founded the Canterbury Museum and was involved in establishing Canterbury University College. A large, jovial and highly cultivated man, Haast married Arthur Dobson's sister Mary. His scientific achievements were recognised by a number of awards, including a knighthood in 1886, a year before his death.

CHAPTER SIX

Across the Alps

Road and rail links

After nearly 150 years of settlement, access to the more remote high-country sheep stations in this area is still by long, narrow and dusty roads. These were developed from rough tracks formed for drays and bullock wagons to take loads of wool out, and even today they require care in negotiating, although oncoming traffic is usually signalled from well away by a plume of dust. Most difficult are the roads to Mount Algidus and Manuka Point Stations, both of which involve risky river crossings. Although there is some 20 kilometres of sealing up the Ashburton Gorge, and a similar amount to Lake Coleridge village, access to the far reaches of the Rakaia and Rangitata Valleys is by gravel road.

A plume of dust heralds the presence of a vehicle on the Mount White road.

On the 'dry weather' road from Lake Lyndon to Lake Coleridge.

This photo, taken in 1887, illustrates the access problems that faced sheep stations in the early years of settlement. These horse-drawn wagons carrying wool bales from Mount White Station would have had to travel more than 85 kilometres over rough tracks, including crossing Porters Pass, before reaching the then railhead at Springfield on the edge of the Canterbury Plains.
A. E. Preece photograph, Canterbury Museum 19XX.2.649

The Waimakariri Valley, by contrast, has been opened up by the construction of the West Coast road (now State Highway 73), and also the Midland railway line, route of the celebrated TranzAlpine scenic rail journey. Both cross the Southern Alps at Arthur's Pass. The story of their building and development is a saga of fortitude, skill and adventure.

The West Coast road

The story of the development of the road that links Christchurch with Kumara, near the mouth of the Taramakua River, begins in 1865 with the discovery of gold on the West Coast, which was at that time part of the province of Canterbury. Gold fever instantly struck Christchurch and within a few weeks the provincial government provided funds for constructing a road across the Southern Alps. The previous year a young surveyor, Arthur Dobson, son of Edward Dobson, the newly appointed provincial engineer, had spent several months exploring and surveying on the West Coast. When crossing Harper Pass from the Hurunui he was told by a Maori chief, Tarapuhi, of a route further south linking the Waimakariri with the West Coast.

Following a request from the chief surveyor, Thomas Cass, to find a pass suitable to carry a road, Arthur and his 16-year-old brother Edward set out up the Waimakariri Valley then up the Bealey. In March 1864 they crossed the Main Divide and scrambled down the other side as far as Otira. This was considered to be very difficult as a route for a road, so a few weeks later Edward Dobson and his eldest brother, George, checked the headwaters of several of the other tributaries of the Waimakariri in search of a better crossing. It is said that when asked by his father which was the most suitable route, George replied, 'Arthur's is the best.' So the pass was named and work began on the road across the Southern Alps.

In 1864 Arthur Dobson discovered the Main Divide pass named after him.
Canterbury Museum 8350

An 1866 sketch by Nicholas Chevalier of the newly completed dray road over Porters Pass.

The offer of good wages attracted a large proportion of the labour force of Christchurch, and within a very short time up to a thousand men were engaged on constructing the road from Porters Pass to Otira and beyond to Hokitika. This was no easy task. Porters Pass rises steeply from the plains to a height of 942 metres (the highest main-road pass in New Zealand), and Arthur's Pass is not only a similar height (920 metres) but is covered with very rough glacial moraine and subject to frequent rain and snow. To compound the difficulties, the winter of 1865 was particularly severe and conditions for those working with picks, shovels and wheelbarrows, and living in tents, must have been miserable. It is astonishing, then, that a dray road was surveyed and built in just one year, and the first coach crossed in 1866.

Ironically, little gold was ever brought across the Main Divide by land to Christchurch, but in the first year after the road over Arthur's Pass was completed it is estimated that about 40,000 sheep and 25,000 cattle were driven over it to feed the multitude of miners seeking their fortune.

The new road also saw the beginning of a twice-weekly coach service by Cobb and Co. between Christchurch and Hokitika. These journeys and their hardships have become the stuff of legend. The trip cost £8 and, even in the best of conditions, took 36

A present-day view of the road climbing towards Porters Pass from the Canterbury Plains.

hours. In 1895 snow was so heavy that coaches were unable to cross the pass for three months. Up to nine passengers would be crammed inside the coach, and eight outside, with those on the roof swaying precariously from a height of nearly three metres. The following description from a tourist brochure suggests the perils faced by travellers:

> The regular river crossings provided special challenges. These braided streams could quickly change from a trickle to a torrent in a few hours. Imagine travelling the road in a poorly sprung coach hanging precariously to the mountain road and only centimetres from a precipitous 100-metre drop into the mountain river below. Your coach would be pulled by steaming, snorting horses. Although the journey uphill was slow (and often passengers had to walk), the journey downhill could be much more exciting, and accompanied by the smell of iron on wood, fast-travelling horses and constant reassurance from the drivers. At each staging post while the ladies sipped tea, most men (including the drivers) found the need for stronger liquid refreshment. This journey, before the days of weather forecasts, would frequently turn into a week-long expedition.

The Reverend Tremayne Curnow conveyed his fear of the journey in verse:

A roadside picnic near the top of Porters Pass in the 1880s.
E. M. Lovell-Smith Collection, Canterbury Museum 1950.80.14

> The view it fairly scares me stiff
> The road's a ledge along a cliff
> The wheels go near it, heavens if
> They make a deviation.
> I only hope that I survive
> This perfect nightmare of a drive
> To offer thanks when I arrive
> Safe at Otira Station.

One writer described a typical coach driver as 'a fine burly man, fearless and dauntless, caring not for

Cobb and Co. coaches on the West Coast road in the Castle Hill Basin about 1904.
James Ring photograph, Canterbury Museum 1950.80.18

Above: Coaches crossing the Waimakariri River in flood in the 1880s.
E. M. Lovell-Smith photograph, Canterbury Museum 1950.80.19

Below: An 1882 photo of coaches outside the Glacier Hotel, Bealey, a major staging post on the journey over the West Coast road.
Burton Brothers photograph, Canterbury Museum 1988.312.3

danger, be it flood, bushrangers or intricate roads hewn out of the forest, beset with boulders and tree stumps: the greater the danger, the better he seemed to like it'. The Cobb and Co. service was run for nearly 50 years by Hugh Cassidy, a large and jovial Irishman.

A number of staging posts along the route allowed for a change of horses and, when required, overnight accommodation for passengers. Inns were built in the upper Kowai Valley before the climb over Porters Pass, at Castle Hill just opposite the present village, at Cass, and near the site of the present hotel at Bealey. In addition, there were staging posts at Springs (near the present turn-off to Porter Heights skifield) and Craigieburn. The original inns are all long gone, but the uncomfortable experience of nights spent in them has lingered on in travellers' tales. Facilities were sometimes overcrowded and meals were generally very basic. One disgruntled guest complained that his dinner consisted of one potato and a piece of steak.

As the Midland railway line was built across the Canterbury Plains, then towards Arthur's Pass, the coach journey was progressively shortened from the extending railhead, finally ceasing in 1923 when the Otira tunnel opened.

With motor vehicles supplanting coaches as the main form of road transport, the West Coast route was

improved, but it was not until 1936 that a bridge was built across the Waimakariri at Bealey. Until then car drivers had to risk fording the river or pay to have their vehicles towed across.

During the 1960s the road was gradually sealed, and although the descent from Arthur's Pass to Otira remained tortuous and prone to rockfall, the highway became a very popular tourist route. With the completion of the Otira viaduct in 1999 the last difficult section was removed.

The building of the West Coast road and its progressive improvement played an enormous part in the development of the Waimakariri Valley. Not only did it make access much easier for those on sheep stations, but eventually it opened up the area to tourist and recreational development of various types. Many claim that the 'Great Alpine Highway' is the finest scenic route in New Zealand. When driving over this road and enjoying the magnificent scenery, it is all too easy to forget the hundreds of men who laboured on its construction, and also the rigours of early coach journeys, with their hazardous river crossings and slides down precarious inclines.

A car stranded in a riverbed near Bealey in about 1910.
J. Anderson photograph, Canterbury Museum 19XX.2.608

The Midland railway line

The story of the construction of the Midland railway line from Springfield through the high country to Otira and the West Coast also saw daunting challenges overcome by skilful engineering and backbreaking labour in a harsh environment. But unlike the road, which was built remarkably quickly, the rail link to the West Coast took over 36 years to complete, during which time the original contractors abandoned the project and then New Zealand's largest construction company was forced to hand over the trouble-plagued job to the Public Works Department.

In 1870 Colonial Treasurer Julius Vogel announced an ambitious plan to build a network of railways to link the scattered settlements of New Zealand. Some track had already been laid in Canterbury, and by 1878 the main trunk between Christchurch and Dunedin was completed, with a number of branch lines extending out into the plains. In 1884 a royal commission decided that the line to Springfield should be continued up the Waimakariri Valley and over Arthur's Pass, instead of the longer but easier route up the Hurunui Valley and over Harper Pass.

In 1887 the New Zealand Midland Railway Company was contracted by the government to build a railway across to the West Coast and on to Nelson, all to be completed within 10 years. This was quite unrealistic in view of the lack of funding and ignorance of the conditions by the company's London directors, and it is not surprising that the venture failed within a decade. About 120 kilometres of track was laid in Westland, but in Canterbury only the Kowai River bridge and a short stretch to Otarama at the entrance to the Waimakariri Gorge had been constructed. After much legal argument, the Crown took over and work was resumed in 1898 by the Public Works Department.

The next section, around the edge of the gorge and across Broken River, was described by the district engineer as 'very rough, the mountain slope rises from the riverbed while the river runs in a fearful gorge all the way'. In just 13 kilometres the line passes through 16 tunnels and crosses four major viaducts. The Staircase Gully viaduct is 73 metres high,

Rail passengers transferring to coaches at Klondyke, the Arthur's Pass station, about 1915, before the Otira tunnel was completed.
James Ring photograph, E. M. Lovell-Smith Collection, Canterbury Museum 1950.80.20

The 73-metre-high Staircase Gully viaduct under construction and in 2005, a hundred years later.
Left: Canterbury Museum 19XX.2.856

A railway enthusiasts' steam train emerging from one of 16 tunnels along the edge of the Waimakariri Gorge and about to cross the Broken River viaduct. The growth of broom and gorse along the route is a typical consequence of early human settlement in the region.
Graham Radcliffe

the loftiest in the South Island, and the Broken River viaduct is almost as tall. With little in the way of machinery to relieve the labour, construction of this section was very slow and difficult. Most of the workers, often with their wives and families, lived in tents pitched on the nearby hillsides.

By 1906 trains were able to run to a temporary terminus at Broken River, where coaches took over to complete the journey to the West Coast in a single day in favourable conditions.

Although the next leg, through to Arthur's Pass village, was comparatively easy, it took eight years to complete. The railway was extended in gentle grades up the Slovens Stream, then skirted a series of lakes before reaching the construction settlement at Cass in 1910. In the next four years the line was taken through a cutting into the Waimakariri Valley, across the river and up the Bealey Valley to Arthur's Pass. Cobb and Co. coaches were now confined to taking passengers across to Otira.

The Main Divide itself now presented the last obstacle. The original proposal was to build a complex rail system over the pass, involving a switchback and rack rail, which would have been expensive and impractical to operate, and was rejected in favour of an even more expensive tunnel. In order to accommodate a major height difference at each end, the longest tunnel (8.5 kilometres) in the southerrn hemisphere at the time was designed to rise from Otira to Arthur's Pass with a gradient of 1 in 33 – the steepest in New Zealand.

John McLean and Sons, a highly regarded construction company, was awarded the contract to build the tunnel and began work in May 1908. Crumbling rock, very cold working conditions, excessive water and persistent labour problems eventually proved too much and in 1912 the company was forced to ask Parliament to release it from the contract with less than half the tunnel completed. Once again the Public Works Department took over and the job proceeded very slowly throughout the war.

During the course of the tunnel project eight men were killed in accidents, and at one stage a rock fall imprisoned 10 men, two of them for three days. When tunnelling crews from each end met in 1918 the surveyors' centre lines were found to vary by only 29 millimetres for line and 19 millimetres for level – a remarkable feat of accuracy given the comparatively crude nature of their instruments. During the next five years the tunnel was enlarged, lined in concrete and electrified before it opened for traffic.

Although steam trains were used on the rest of the line, electric locomotives took over through the tunnel because of the steep gradient, and the fact that the heaviest loads (coal) would be coming up the slope from Otira, where steam engines would present problems with smoke and furnace gases. The six locomotives that shuttled through the tunnel were powered by a coal-fired electricity generating plant at Otira.

In August 1923, after 36 years of construction, the 63-kilometre-long Midland line between Springfield and Otira was completed. The only railway to cross the Southern Alps, it has gained a reputation as being one of the world's great train journeys.

Kb Country

The story of the Midland line would not be complete without a look at the trains themselves, and especially the locomotives. The main use of this railway has always been to transport bulky goods – timber and particularly coal – from the West Coast to Christchurch and its port at Lyttelton. The combination of heavy loads and a number of relatively steep grades on the line meant that powerful engines were required.

Six Kb steam locomotives, constructed mainly for this route, began service in 1939. They were based in Springfield, and the area between that town and Arthur's Pass soon became known by train buffs as 'Kb Country'. These massive engines were a spectacular sight as they belched great clouds of steam while

A sight to warm any railway buff's heart, this excursion train is pulled by one of the Kb locomotives that plied the West Coast railway route from 1939 until 1969.

Courtesy Jack Creber

A Kb engine hauling a load of wagons from the West Coast and working hard on the climb up Cass Bank, near Lake Sarah, the steepest section of the line between Arthur's Pass and Springfield.
Courtesy Jack Creber

The TranzAlpine Express leaving the Waimakariri Valley and turning towards Cass enroute from Greymouth to Christchurch.
Graham Radcliffe

grinding their way up inclines in the heart of the high country. The Kb locomotives were the most powerful in New Zealand, weighing 146 tonnes and capable of pulling loads of up to 550 tonnes up grades of 1:50.

Most famous of all the engines was Kb968 – 'The Mighty Cannonball' – commemorated in a ballad written by John Cooke:

> . . . From the Pass to Broken River,
> Through sun and sleet and hail,
> Up through tunnel number six –
> On wet and greasy rail.
> Six hundred tons behind her
> But seldom would she stall
> While heading down to Springfield on
> The Mighty Cannonball . . .

The Springfield to Arthur's Pass line was isolated, and stories about the antics of train crews are numerous. It was not uncommon for some to carry rifles and shoot deer from the engine. On one occasion the staff at Springfield became alarmed at the non-arrival of a goods train. When it finally arrived in the small hours of the morning, the crew blamed the delay on problems with the coal. It was later revealed that the train

The 'Perishable'

The 'Perishable' was a nickname given to the train that travelled between Christchurch and Greymouth six nights a week carrying perishable goods. Although it did not normally take passengers, it was well known to mountaineers and trampers (including myself) because it provided a means of getting a weekend's recreation in the Arthur's Pass region.

On a Friday evening, a group of people would assemble in the Christchurch railway goods yard and, with permission, we were allowed to travel in the guard's van. After much shunting and an uncomfortable journey, the train would reach the upper Waimakariri Valley around midnight and then slow down at various spots to drop trampers off.

I have vivid memories of leaping out as the 'Perishable' negotiated the curve at Klondyke Corner. Our small group huddled in the dark, watching the train's light disappear up the Bealey Valley and knowing we had a long slog ahead of us before we would reach one of the huts up the Waimakariri.

Arthur's Pass from the head of the Bealey in 1865, showing an early campsite.

Ward A. Reeves lithograph from a sketch by Edward Dobson, Canterbury Pilgrims and Early Settlers Association Collection, Canterbury Museum 1948.148.294

had been shunted onto an isolated siding while the crew went deer shooting.

There were other, more dangerous incidents, including a little-known report of a train with an full load of coal wagons racing at high speed through the tunnels around the Broken River area as a result of brake failure. The Kbs were a tight fit in the tunnels, and sometimes the engine men wrapped wet towels round their heads to cope with the smoke.

In 1969 the era of steam trains came to an end, and the Kbs were replaced with more powerful and less labour-intensive diesel-electric locomotives. Even the electric engines used in the Otira tunnel were replaced by diesel locomotives, in 1997, when new ventilating systems were installed.

Passenger use of the Midland line was given a boost in 1987 with the introduction of the Tranz-Alpine Express. This train, with its large observation windows, upgraded seating and onboard catering, has become the most successful passenger rail service in New Zealand and provides an ideal way of viewing the spectacular high-country scenery in comfort.

Arthur's Pass Village

The settlement of Arthur's Pass began as a camp for road builders and was originally known as Camping or Bealey Flat. In the first decade of the twentieth century it became home to the builders of the Otira rail tunnel, and a number of workers' huts, dining rooms and machinery sheds were erected.

Because of the large number of transient labourers earning relatively high wages, and the ready availability

Arthur's Pass in about 1910, looking south from Wardens Creek, at a time when the settlement housed railway construction workers.

James Brake Collection, Canterbury Museum 1983.318.2

This painting of Arthur's Pass by Louise Henderson depicts the little village in the 1940s.
Alexander Turnbull Library G-685

of explosives needed for construction, Arthur's Pass developed a reputation for drunkenness, unruly behaviour and violence. But life there gradually settled down and, once the tunnel was completed in 1923, the character of the village began to change.

Many of the railway houses became holiday homes for Christchurch people with a love of the mountains. A number of new baches were also built and it was decided that there would be no fences and any plantings should fit in with the surrounding bush.

In 1929 the status of the village was considerably enhanced by two events. The first was the arrival of Oscar Coberger, who introduced the sport of skiing (see page 126), and the second was the announcement that the surrounding area was to become a national park. Part of this had been recognised as a reserve in 1901, following the initiative of botanist Leonard Cockayne, but once full park status was declared, interest in the area grew rapidly (see page 147).

Several tramping clubs built huts and hostels and Arthur's Pass became a hub for mountain sports in Canterbury. Later, as money became available, rangers were appointed and a national park headquarters and visitor centre was established.

Arthur's Pass now has a store, restaurants and a range of accommodation catering for tourists driving through. In recent times skiing has become a less important drawcard as other, larger skifields have been built. New holiday homes have tended to cluster at Bealey Spur, which, just 13 kilometres down the valley, is a much sunnier and drier spot. Thus Arthur's Pass remains small and, to some eyes, perhaps a little run down, but it retains its character as New Zealand's only European-style alpine village.

CHAPTER SEVEN

Where Wool is King

The era of the big sheep runs

Wool has dominated the lives of most of those who live in the high country of central Canterbury. The first European settlers came to find pasture suitable for the rearing of sheep and hoped to make a fortune from the sale of wool. By 1860 almost all of the land had been taken up and subdivided into about 30 pastoral leases. Nearly 150 years later, although the area of farmed land has shrunk considerably, there are about 45 runs, most of which continue to rely to a greater or lesser extent on sheep as their main economic base.

As we have seen, the first 50 years of farming in the high country saw mixed fortunes for runholders, but successful management patterns gradually emerged and, by the 1920s and '30s, most stations were operating in a relatively stable and profitable fashion. In fact, a romance was developing about life on the big runs, celebrating hardy musterers and their legendary dogs, together with itinerant shearing gangs and, of course, the iconic merino sheep. This perception of the high country was popularised by a number of books and photographic essays inspired by the dramatic scenery of the region and the quirky independence of its inhabitants.

From about 25 years ago dramatic changes have been taking place in high-country land management, as growing wool is being displaced as the predominant activity. The following descriptions are of a farming system that is now very much in transition for most of the properties in this region, particularly as leasehold properties complete tenure review.

The traditional sheep run

The sheep runs of the Canterbury high country are generally centred on a homestead, located on freehold sheltered valley flats. Surrounding this is an area of hill country, leased from the government for periods of 33 years with right of renewal. Particularly

Merino sheep waiting to be put onto winter feed on Mount Arrowsmith Station, near Lake Heron.

in the more rugged highlands, these properties are often very large – Mount Posession/Hakatere comprises about 27,000 hectares, even with some 17,000 hectares of the most difficult country retired from grazing. Mesopotamia, on the south side of the Rangitata, and Mount White, in the upper Waimakariri, are of similar size. Most of the runs in the lower hill country to the east are privately owned and considerably smaller, several being only 1,000 hectares.

The value of the more elevated high country or grazing depends very much on its physical features. Accessible and sunny faces, where snow does not lie and with good natural boundaries, can be stocked for most of the year. Much of this country, though, was used only for summer grazing and has recently been retired from stocking altogether.

The key to the viability of a sheep run largely depends upon the amount and quality of its lower, more sheltered freehold land. The extent to which this can be improved by topdressing or cultivated to provide winter feed largely determines how much stock can be carried through the winter.

The homestead

Each of the five roads up the major river valleys runs past half a dozen sheep stations, each several kilometres from its nearest neighbours. The focal point is the homestead and attendant buildings, whose presence is signalled by clusters of mature, mainly coniferous trees. Some homesteads are more than a hundred years old, and in several cases are very impressive buildings. The original homestead at Grasmere was built in 1857, and a portion of this has survived. More commonly, a new homestead has been added on to or completely replaced the old.

The huddle of other buildings on the run will be dominated by a woolshed, usually of corrugated iron, large enough to accommodate well over a thousand sheep, and surrounded by a network of yards. There will be housing for permanent employees, shearers' quarters with a cookhouse, stables, machinery sheds, perhaps a hay barn and a killing shed. Not too far away will be kennels for the many dogs that are still an indispensable part of sheep management. Several stations now boast an aircraft hangar.

Merinos and halfbreds

The first pastoralists experimented with several breeds of sheep, but most settled quite quickly on the merino, brought over from Australia. This breed produces very fine wool and, most importantly, can endure extremes of heat and cold. Because merinos tend to be less docile than other breeds, it is easier for dogs to get them moving on hill country.

The breed's disadvantages became apparent when the arrival of refrigerated ships created a market for frozen mutton in Britain. Merinos have a relatively low lambing percentage and they do not produce good meat, as any musterer or shearer who has lived on a diet of tough chops will testify. Although some stations subject to the harshest of winter conditions continued to run pure merinos, many pastoralists began to cross their ewes with rams of good mutton-producing breeds such as the Leicester. The lesser

The homestead area of Ben McLeod Station, in the middle Rangitata Valley, is typical of many high-country sheep runs. It is surrounded by a cluster of buildings, including a large woolshed, stables, implement sheds and employees' quarters, framed by stands of mature exotic trees.

Cattle being moved at Mount Potts Station, with the mountains in the headwaters of the Rangitata in the background.

value of the wool from these halfbreds, as this cross became known, was generally more than compensated for by the farmers' ability to sell surplus stock. A recent innovation has been the introduction of small flocks of sheep bred primarily for meat production.

The success of high-country farming is dependent on wool prices, and these fluctuate widely. Perhaps the heyday of pastoralism coincided with the very high prices prevalent in the 1950s. Recently, a depressed market for wool has forced some farmers to diversify by running beef cattle such as Hereford or Angus on suitable land. These range from herds of 200–300 through to the 3,500 head on Mount Possession/Hakatere.

Further diversification was achieved in the early 1970s by the Urquhart family of Erewhon Station, who built up a herd of deer by capturing and breeding wild animals. A number of high-country properties have now erected tall fencing and farm deer to produce venison and antlers. Some runs have more than 2,000 head, and Mount Hutt Station is stocked with nothing but deer.

It should not be forgotten, however, that wool is still king in the high country. There are about 300,000 sheep on the 45 runs in central Canterbury, and they represent close to 80 per cent of stock units.

The sheep-run year

For the past hundred years or so, the activities involved in sheep management on a large high-country run have changed little. They are very seasonal, ranging from quiet periods during the winter, when perhaps only two or three permanent staff may be employed, to the frantic activity surrounding shearing, when 15 or more additional people are required.

The autumn muster brings sheep down to lower and sunnier country, then stock are moved onto winter feed on the flats. After heavy snowfalls, some strenuous snow-raking to rescue stranded sheep may be required. Ewes are kept on lower country for spring lambing, after which tailing and dipping are carried out.

Mustering has become evocative of the way of life on a high-country sheep run, and has been lyrically described by writers such as David McLeod, Peter Newton and, more recently, R. M. Burdon and others. Stories are told of musterers arriving on the top beat with the mist still filling the valley below, of the amazing feats of sheepdogs, of stubborn stock that would not move . . .

These images have lingered, despite a good deal of the more inaccessible country having been retired, four-wheel-drive vehicles and farm bikes replacing horses on the easier country and, on big stations such as Mesopotamia, the use of a light plane or helicopter in the muster.

An equally distinctive aspect of high-country farming is concentrated upon the frantic few weeks associated with shearing, which generally takes place in late spring and early summer. This task still involves the arrival of a gang of perhaps five to eight shearers and a similar number of shedhands to pick up the fleeces, keep the floor swept and operate the wool press; a wool classer, who is in charge of the shed; and the all-important cook. During this time the weather becomes critical, as it is important to have a thousand or more sheep dry for shearing each day. A very

Above: Sheep dipping in 1899 on Orari Station on the southern margin of the Rangitata.

F. Bradley & Co photograph, Canterbury Museum 19XX.2.855

Right: Modern-day dipping, often carried out by contractors using a mobile device, illustrates one of the ways that high-country farming is changing.

Below: Drafting lambs on Ryton Station.

large and busy shearing shed holds a special fascination for the 'townie', assailing all the senses with unusual sights, sounds and smells.

Machine shearing became common in the 1940s, but many high-country runs have continued to use blades. Although slower, this practice leaves more wool on the sheep and enables them to withstand cold conditions. The trucking out of loads of wool bales brings to an end the year's work, the success of which then hinges on fickle market prices.

Dipping is another labour-intensive activity on most sheep runs. This involves forcing stock through a bath containing a strong pesticide to eliminate parasites. It was a particularly important job in the earlier years of settlement when scabies was rampant. Later, sheep ticks, or 'keds', were treated in this manner.

Other sheep-focused activities include crutching, moving of stock, sorting out and selling surplus sheep, and cultivating the flats for growing winter feed. On many properties these traditional tasks are now combined with managing cattle, deer and, increasingly, tourists.

Although it has been said that the musterer and his dog are being replaced by the Toyota and helicopter, the sheepdog remains a key ingredient in day-to-day activities on the sheep run. A well-bred and trained dog is a valuable asset and most shepherds will have a pack of at least three or four, including a good huntaway. Bred originally from border collies specifically for New Zealand conditions, these are strong, vocal, short-haired dogs that will willingly drive mobs of sheep wherever required all day. Another valuable helpmate, particularly for working skittery merinos on hills, is the heading dog, familiar as the stars of televised dog trials. These can turn sheep, often at great distances, and quietly hold or move them where required. Good additions to the pack are dogs trained to move sheep in the yards, and, even better, multi-tasking 'handy' dogs.

The people of the high country

Much has been written about the colourful characters who have worked in the high country – musterers, shearers and rouseabouts – and their horses and hard-working dogs. But it is the runholders and managers who have been the real key to the operation of the sheep stations. They have weathered bad times and good, and have introduced the changes necessary for the survival of pastoral farming.

In the early days few sheep runs remained in the same hands for very long, but as more stable ownership patterns developed, exceptional individuals began to emerge as leaders in high-country farming circles.

Keeping a keen eye on sheep in the yards.

Robert Todhunter, who served as the first chairman of the High Country Committee of Federated Farmers, was noted for the quality of his merino flock at Blackford at the start of the twentieth century. In 1917 he took over the remote and difficult Lake Heron Station, where a fourth-generation Todhunter, Philip, continues to farm.

David McLeod ran Grasmere Station in the Waimakariri for 40 years from 1930 and was succeeded by his son. He also chaired the High Country Committee and the Tussock Grasslands and Mountain Lands Institute for many years. His books and other publications in the 1960s and '70s did much to publicise both the romance and the problems of pastoral farming.

Alex Urquhart was at Lake Heron in the early days, and his sons managed several stations, including Mount Algidus and Mesopotamia. A grandson, Arthur, took over the lease of Erewhon in 1943, and he and his son farmed there for 53 years. The Urquharts' enterprise resulted in their pioneering deer farming using captured wild animals, and they also established a safari hunting operation.

Another long-term dynasty began in 1931 when Leo Chapman started developing what was to become

Inverary Station, a property now in the hands of his son John.

We have already met (see pages 73–74) the remarkable John Acland and learnt of his family's occupation of Mount Peel from 1857 to the present day.

Among the renowned characters of the high country was Sam Chaffey, who managed Mount Possession for over 30 years. He was not only a great stockman and innovative farmer, but was the source of a great fund of stories. In *The Stations of the Ashburton Gorge*, John Chapman writes:

> On one trip down the Ashburton Gorge Road, Sam ran out of petrol in a howling nor-west wind. He claimed he made it right down to Mt Somers by tying the car doors open and 'sailing' down. Never one to let the truth stand in the way of a good story, he added that 'when I went round a corner I just closed the door'.

While there are similarities in the appearance and operation of the 45 or so central Canterbury high-country sheep stations, each has its own character. In the following pages these runs are identified, and in some cases briefly discussed, in sequence along the roads up the major river valleys.

Waimakariri Valley sheep stations

State Highway 73 from Christchurch to the West Coast passes through all but two of the runs in the Waimakariri Valley.

After reaching Springfield the highway turns left into the Kowai River valley and enters an area known as the 'front country', which includes three runs. **Torlesse** Station is to the right, covering the country at the eastern end of the Torlesse Range as far as the rugged bluffs along the Waimakariri Gorge. Established very early, it is still largely running sheep, though the owners also host parties of tourists.

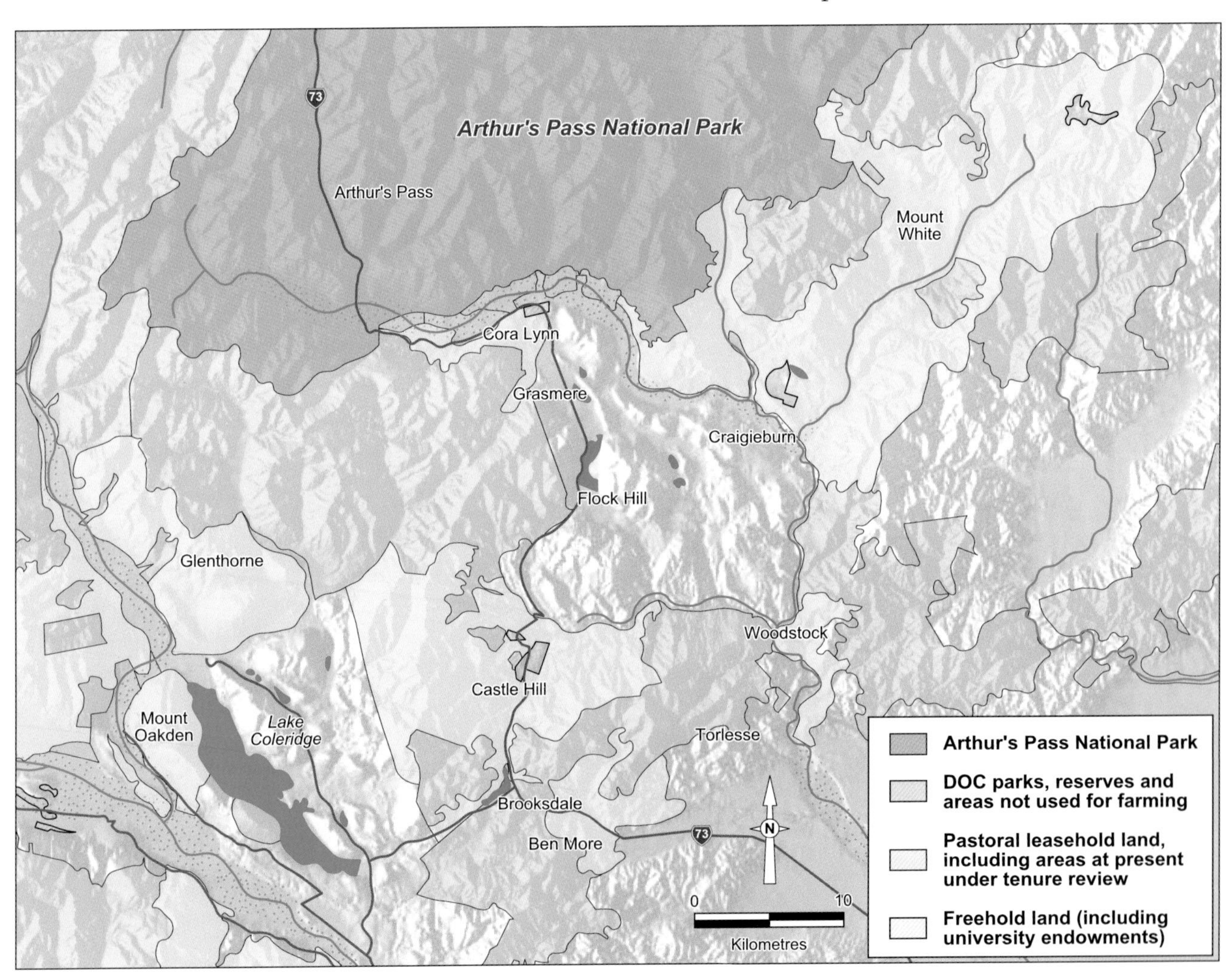

Waimakariri high-country sheep runs.

Cartography by Tim Nolan

On the left (west) side of Mt Torlesse toward Porters Pass lies **Brooksdale**, and on the south side of the Kowai River is **Ben More**, which extends up into the Big Ben Range. Recently, most of the higher country on both runs was taken over by the Department of Conservation and incorporated in the Korowai/ Torlesse Tussocklands Park. In 2002–03 each of these reduced properties was purchased by overseas interests, both of whom announced major changes to the runs' traditional wool-growing operations. Brooksdale was to be farmed in a much more intensive and varied way, involving the rearing of quality sheep and beef cattle, with forestry on the hills and seed production on the flats. The new owner of Ben More planned to develop a safari hunting park.

Once over Porters Pass SH73 heads towards a number of big sheep runs in the upper Waimakariri Valley. First to appear on the left is **Castle Hill**, with its spectacular limestone outcrops. (Some of the limestone for the Christchurch cathedral came from this site.) This property was taken up by the Porter brothers in 1858. Six years later it passed to the Enys brothers, who built a homestead, Trelissick, near the present Castle Hill village (see page 79). This has not survived, but another homestead, built of limestone in the 1860s on the site of the present house, is being converted to provide visitor accommodation. The diaries of John Enys are a good record of high-country life from 1864 to 1890. He took a strong interest in local affairs and was a well-known naturalist and keen fisherman.

Most of this station is over 700 metres above sea level, and the homestead is the most elevated in our region. Although winter conditions are severe and the growing season for feed is short, Castle Hill has been farmed continuously for nearly 150 years. In 2004 the government, using $3.5 million from the Nature Heritage Fund, bought two-thirds of the 11,000-hectare property to add to surrounding conservation parks on the Craigieburn and Torlesse Ranges. The remaining lower part of the station was bought by an Auckland businesswoman, Christine Fernyhough, who is upgrading the property by improving the yards and buildings, increasing cultivation for winter feed and upgrading the quality of the livestock (see page 102). At present the run is stocked with about 4,000 merinos, 350 cattle and 250 red deer.

After leaving Castle Hill, the highway descends through the Craigieburn cutting and, before reaching Lake Pearson, passes **Flock Hill** on the right. This run now boasts a lodge with a restaurant and an extensive range of accommodation, but still operates as a substantial farm, running sheep, cattle and deer on 14,000 hectares. Flock Hill was originally part of the huge block of land taken up in 1857 by Joseph Hawdon. This also included what are now Grasmere and Craigieburn Stations. Flock Hill become a separate run in 1917.

The old Castle Hill limestone homestead and farm buildings on a winter morning. The Craigieburn Range looms in the background.

A short distance beyond Lake Pearson, and on the left of the main road, **Grasmere** Station homestead nestles below Mt Misery and is surrounded by trees. This is one of the best known of the region's sheep runs and the story of its past and recent changes are typical of the Canterbury high country.

Grasmere was part of the large area taken up by Joseph Hawdon in 1857. The original homestead was built of cob (mud and straw bricks) and slab timber in 1858, and later faced with Castle Hill limestone. In 1903 a wooden wing with verandas was added, and this building forms the core of the present house.

The property changed hands several times and was run in various combinations with neighbouring stations until in 1930, at a time of falling wool prices, it was sold to David McLeod. He spent the next 40 years there before handing it on to his son.

Since 1988 the owners of the freehold of Grasmere have continued to run 3,000 merino sheep, as well as deer and beef cattle, on 5,000 hectares, but tourism has become the main enterprise. Grasmere Lodge is promoted as a luxury retreat and offers a range of high-country activities.

Nobody said it would be easy!

In 2004 Christine Fernyhough bought the remains of the Castle Hill run after it had been subdivided and the Department of Conservation had taken over much of the higher part. Her views are of particular interest because she came from Auckland and had no previous experience of high-country farming.

Her forthright and perceptive comments paint an at-times painfully realistic view of the joys and problems of trying to run this challenging property. Below are exerpts from an address she delivered a short time after the big snow in June 2006.

'When I bought and moved to Castle Hill people said, "Amazing, how brave, I admire you!" – but others knew that Castle Hill is very difficult to farm.

'As I'm finding out, it is a mite more difficult than it was when viewed from afar. I used to wonder how farmers could be happy with a four per cent return on capital. Now I'd give my all for that rate of return. The market has moved downwards in every stock line I have – wool, deer, beef and sheep – since I purchased . . .

'They say farming is all about nutrition, nutrition, nutrition and genetics. But when you fall in love with a slice of iconic high country between 700 and 1,000 metres above sea level, in the heart of the Southern Alps, when you are bowled over by its beauty, the size of the sky, the rocks, the tussocks, you fail to see the state of the fencing, the cost to revamp a newly configured farm and, more importantly, the depth of topsoil – the potential to grow grass!

Christine Fernyhough.
Courtesy John Bougen

We have sweet limestone country but grass and winter feed is hard to produce with such a short growing season before the sun turns the landscape the colour of toast and the ground crackles under your feet. Winter means 130 days of feeding out, changing break fences in five degrees below with a wind chill of whatever . . .

'We're still covered in snow. Except for a small area of melt on the northern slopes, our stock haven't had grass beneath their feet for a month . . .

'It's all about seasons, managing cycles, being alert to the weather, to pasture growth, worms, teeth, foot rot, the serendipity of mating and then the harvest – lambs and calves – flourishing . . .

'Farmers live in the hope that their hard work – so often starting with the letter "d": drenching, drafting, dipping, dagging, docking, debt and, in my case, dealing with DOC – will lead to better returns. Farming is the only business where everything is bought at retail, sold at wholesale and we pay the freight both ways!.

'One of the best things about my move to Castle Hill has been the wonderful high-country people who have become friends. Because I'm prepared to give anything a go, they've been encouraging and quick to share their knowledge and experience.'

Castle Hill after the June 2006 snowfall.
Courtesy John Bougen

Left: Grasmere Station buildings shortly after construction in 1858. *Right:* The homestead a few years later.
Below: Part of the original homestead can be seen in this present-day photo.

A short distance beyond Grasmere a side road runs to the right, following the railway line past Lake Sarah and down the Slovens Stream valley. About 12 kilometres along this side road is the homestead of **Craigieburn** Station, which now extends over mainly moderate hill country between the Waimakariri River and Flock Hill but has had various boundaries and was once much larger. In 1917 the run was bought by Walter McAlpine, who ran it for some years before his son John (later a Cabinet minister) took over. The present owner continues to operate Craigieburn as a pastoral property but is also following the trend of expanding into tourism by operating Alpine Safaris, a game-hunting and fishing-guide business.

Back on the highway, a group of mature pine trees marks Cass, once a thriving railway settlement but now almost deserted (sse page 153). The surrounding land is owned by the University of Canterbury; on the hill behind the old rail depot is a biology field station. Most of Flock Hill and Craigieburn Stations is leased from the university.

As SH73 crosses a ridge and drops down to the Waimakariri River, the road to **Mount White** Station branches off to the right. This crosses the river, then winds for nearly 30 kilometres down the north side of the river before reaching the homestead. Mount White is one of the largest sheep runs in the region. Comparatively isolated and containing much rugged

This view of Mount White Station emphasises its isolation. The homstead is at the end of a 30-kilometre-long gravel road and at least 35 kilometres from its neighbours.

country, it has had a chequered history, being combined with other properties at times and changing hands frequently (see page 82 for details of its early years). It was famous early on for breeding Clydesdale horses, and later achieved distinction for its beef cattle. Mount White was recognised by musterers as a tough run to work on but a good place on which to obtain experience. It continues to operate in a traditional pastoral fashion. Peter Newton, who managed the station in the 1950s, publicised the romance of the life of the musterer and his dogs in *The Boss's Story*.

The last of the Waimakariri sheep runs bordering the highway is situated on the left before reaching the Bealey Hotel. This is part of the old **Cora Lynn** Station, which now runs a flock of merinos on its remaining 2,400 hectares of grassland. The property combines a sheep station and nature reserve. The major operation on the property is an ecotourist venture, Wilderness Lodge Arthur's Pass. This accommodates up to 40 guests, attracted by a guided 'nature discovery' programme.

Rakaia Valley sheep stations

Roads up each side of the Rakaia River provide access to half a dozen large and well-known sheep stations. Before reaching these, the route up the northeast side passes a number of smaller properties on the river terraces, and side roads run to several front-country runs between the Malvern Hills and Big Ben Range.

Beyond the little school at Windwhistle on SH 72, the road leading to Lake Coleridge passes over a series of terraces and moraine hills. The make-up of properties on this relatively gentle country has changed dramatically in the last hundred years or so. Initially, almost all of this land on the north side of the Rakaia was taken up as part of the Snowdon and Acheron Bank runs. Bayfields was divided from the former and, further up the valley the hillier country of the latter became Mount Oakden and Peak Hill.

A major subdivision took place after the Second World War when the government bought much of this land for soldier settlement. The area now consists of about a dozen relatively small freehold properties that benefited from cheap government loans in the 1970s, and have been well developed by topdressing and cultivation. They contrast markedly with the traditional runs further up the Rakaia Gorge, and their exposure to ferocious winds has been reduced by a succession of shelterbelts.

Among the first properties encountered on the left along the Coleridge road are two examples of innovations in the high country: **Terrace Downs** has become a luxury golf resort and **Tui Creek** is to be developed as a series of lifestyle blocks. Both were part of the **Bayfields** Station, which, with **Ben Lea** and **Fighting Hill** on the north side of the road, was subdivided from Snowdon.

As the road continues towards Lake Coleridge, it

passes through **Middlerock**, **Dry Acheron**, **Big Ben**, **Acheron Bank**, **Lake Hill** and **Coleridge Downs**, all of which have been transformed in the past 50 years from windswept tussock terraces into attractive farms.

Further up the road, beyond the power station, is **Peak Hill**, the easier country of which has been devoted to intensive stock farming while the higher areas have been passed to the Department of Conservation as a result of tenure review. **Mount Oakden**, at the head of the lake, sprawls across mostly steep, difficult land.

Bayfields homestead area with Mt Hutt in the background.

Back at Terrace Downs, a loop road leads around Fighting Hill and provides access to High Peak and Snowdon Stations, which once included all of the hill country running into the Big Ben Range and the Malvern Hills.

Snowdon was the original station in the area, occupied as early as 1853 as several runs and amalgamated the following year by Francis Leach, a Welshman who was soon involved in explorations with Thomas Potts. In a key position at the entrance to the Rakaia Valley,

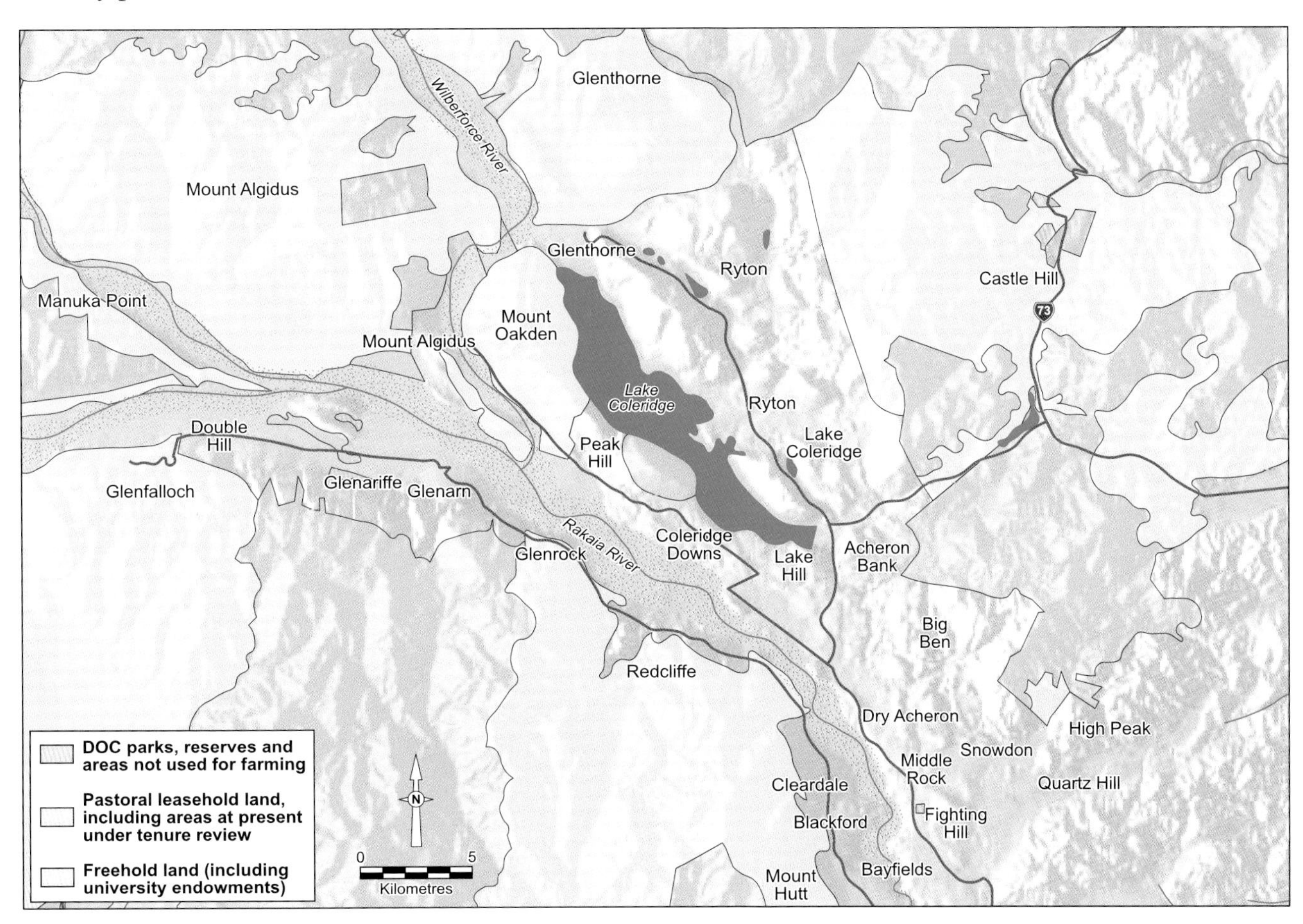

Sheep runs in the Rakaia Valley.

Cartography by Tim Nolan

Snowdon homestead, built by William Gerard in the 1920s, is surrounded by trees, many planted by Francis Leach, who owned the run from 1856.

Snowdon was a popular stopping place for gold prospectors following the early route past Lake Lyndon to the West Coast. The run was eventually bought by George Gerard, who also owned Mount Algidus and Double Hill, and remained in the hands of that family until 1946, when it was sold, reduced in size and the lower country extensively developed. The homestead area was heavily planted in trees, especially by Leach, and several fine homesteads were built on the site. The second burned down in the 1920s and was replaced by the present imposing house.

High Peak Station, too, has undergone considerable change since it was established in the 1850s. The original property extended between the bush-covered hills of the Big Ben Range to the north and the Malvern Hills to the south, including the headwaters of the Selwyn River. Its early settlement is a little unclear, though it is known that the first homestead was built in 1879. In 1884 it was bought by George Rutherford, owner of the nearby Ben More and Dalethorpe runs. High Peak was quite small, with a little under 7,000 hectares of relatively easy land. Like other foothills runs, it occasionally suffers from heavy snowfalls.

Rutherford died in 1918 and the station was inherited by his daughter. It continued to be run by a manager until her death in 1948, when other family interests took over. In 1951 it was partly subdivided for soldier settlement, and in 1980 the remainder was split into two properties, run by the Guild brothers. **Quartz Hill** includes the area to the south and west, and High Peak the central valley and hills to the northeast. Under enlightened management, both have

Looking across High Peak Basin towards Quartz Hill Station, which incorporates the background hills and terraces to the right. Fifty years ago this area was covered in tussock grassland, except for part of the valley floor and some young shelterbelts.

benefited from considerable investment and been extensively developed, with the terraces cultivated and many trees planted. The two properties have more than trebled their stock-carrying capacity and run deer as well as sheep. High Peak also operates a safari park for trophy hunters (see page 136).

Turning right off the road to the power house at 'Dog Box Corner', another road leads to **Lake Coleridge** Station. This was taken up in 1855 and for most of the next hundred years was a very large run, covering over 30,000 hectares between the Rakaia River and the Craigieburn Range and from the Acheron River inland to the Harper, and the very remote Glenthorne up the Wilberforce Valley. In 1890 the station was taken over by John Murchison and parts of it continued to be owned and operated by members of his family until 2001. Over the past 45 years the property has been progressively reduced in size. Much of the high land was retired from stocking and, in 1960, the Glenthorne and Coleridge Downs blocks were sold. In 1983 Ryton was divided off, leaving the present Lake Coleridge Station a remnant of what had been a huge area.

Just beyond the Coleridge homestead a road to the left leads along the north side of the lake to the Harper River intake, passing through rolling moraine country and a series of steep, isolated mountains. About 10 kilometres along this road the cultivated land and buildings of **Ryton** Station appear. This property comprises nearly 15,000 hectares, most of it leased from the Crown and the University of Canterbury.

Although the property runs 8,000 merino sheep, its owners are also at the forefront of change in the Rakaia Valley, operating an expanding tourism venture (see page 132).

About another 10 kilometres beyond the Ryton River, the road crosses the Harper River and continues to **Glenthorne** Station, which takes in the country between the Wilberforce and Avoca Rivers. This remote and difficult property was taken up as early as 1858. After being incorporated into Lake Coleridge, it has operated as a separate concern for over 30 years. Its owners run sheep and cattle, and also provide tourist accommodation.

The legendary **Mount Algidus** Station is situated between the Wilberforce and Rakaia Rivers and is reached by travelling nearly 30 kilometres beyond the Lake Coleridge power station. Before you reach the homestead, the Wilberforce must be forded, a task

A watercolour of High Peak homestead, built in 1879 by Sir John Cracroft Wilson for his manager. Its solid concrete walls were over 30 centimetres thick.

Courtesy Janet Holm

The homestead after the heavy snowfall of 1939. The building is little changed but is surrounded by mature trees.

This 2005 photograph shows a house considerably modified by the Guild family, with a second storey added and extensive gardens planted.

that can be impossible during floods, which have been known to isolate the property for months on end.

This notorious river crossing has swept riders off horses, carried away many vehicles, drowned at least one person and buried another in quicksand. All manner of transport has been used to conquer the river, including bulldozers, heavy trucks, a jet-boat and, recently, aircraft, but it remains to this day a dominating aspect of life on the station. The wife of one of the early runholders found the conditions so discouraging that she refused to remain on the property, but most who have lived there have relished the challenges. The rewards and tribulations of existence at Mount Algidus are the subject of Mona Anderson's much-loved book *A River Rules My Life*.

This run does have some large flats that carry stock, but much of the country is too rugged for economic farming. High-quality wild deer herds in the area attracted the attention of hunters in the 1920s to the '40s, and the many kea that once inhabited the surrounding peaks were considered such a threat to sheep that they were shot in large numbers.

A narrow gravel road runs for almost 50 kilometres along the base of the Mt Hutt, Black Hill and Palmer Ranges, providing access for eight runs south of the Rakaia River.

Mount Hutt Station, which extends across the plains around the base of the mountain, was taken up early in the 1850s and for a time was combined with the adjoining Blackford run. Unusually for this type of country, dairying was once the major activity at Mount Hutt. Early in the twentieth century it carried 600 cows, and a dairy factory (now a museum) stood at the turn-off to the Blackford road. By the 1960s, sheep reigned supreme and a number of shearing records were established at Mount Hutt. More recently, it has been stocked exclusively with about 15,000 deer.

The first two properties along the Double Hill road are **Blackford** and **Cleardale**, occupying the Rakaia terraces and fans on the side of Mt Hutt. Both have been extensively developed and Cleardale carries as many as 10,000 sheep. It is run by Ben Todhunter, who has followed his ancestor Robert as chairman of the High Country Committee of Federated Farmers.

Over the next 40 kilometres the now-gravel road passes the homesteads of five more properties – **Redcliffe**, **Glenrock**, **Glenarn**, **Glenariffe** and **Double Hill** – before reaching Glenfalloch. These properties were subdivided from the original Double Hill run, which was purchased in 1916 by Hugh Ensor, and some are occupied by members of his family.

All are now well-established farms, having considerably improved the lower freehold land and retired the higher country, most of which is currently involved in tenure review.

Mount Hutt Station in winter sunshine. Built in 1880, the two-storey homestead for many years provided homestay accommodation.

Downsizing in the Rakaia

The history of Double Hill Station encapsulates the evolution of land management on the larger high-country runs. Much of this vast run was taken up in 1858, and by 1874 it included all the sunny, north-facing slopes of the Mt Hutt, Black and Palmer Ranges, plus Manuka Point, with some 40,000 sheep grazing 80,000 hectares. A major innovation before the end of the nineteenth century was the construction of a 25-kilometre-long snow fence that divided the high 'summer' country from the lower 'winter' country. During this time Double Hill was run by William Gerard, followed by his son George. It was extremely isolated, with no road access to the homestead until 1925.

The original cob cottage built about the 1860s at Double Hill.
Courtesy Tim and Anna Hutchinson

In 1912 the property was split into four runs, and Hugh Ensor purchased three of these – Glenrock, Glenariffe and Double Hill – in 1916. They were largely run as a unit, and in 1930 employed 11 staff including three rabbiters. After Ensor's death in 1943, his son Peter took over Double Hill and other brothers inherited Glenrock and Glenariffe.

The highland areas of these runs are steep with much shingle, but the lower fans along the base of the mountains are relatively good country. From the 1950s a great deal of development has been carried out, including extensive fencing, the establishment of stock-watering systems and the growing of winter feed, while aerial topdressing has improved the pasture. In 1970 more than half of the property was retired from grazing. Double Hill has recently completed tenure review and most of the higher land is now administered by DOC.

In the past 80-odd years the station has reduced its area by 80 per cent (from 40,000 to 7,900 hectares), yet its sheep numbers have dropped by only 30 per cent (17,000 to 11,500) and the amount of wool per sheep has doubled to 4.8 kilograms.

The present homestead is tucked into the apex of an old fan and surrounded by mature trees and an attractive garden.

Looking across Glenfalloch Station to the Arrowsmith Range.

The last run on the road, **Glenfalloch**, backs onto the Upper Lake Heron Station and is run by another member of the Todhunter family, who combines farming with a heli-skiing operation in conjunction with Mount Hutt Helicopters.

Across the wide bed of the upper Rakaia is **Manuka Point** Station. This remote property was not stocked until 1864 and at times (including the present) has not been farmed at all. It is steep, contains a great deal of scrub and was once described, perhaps by a disgruntled musterer, as 'the roughest shop in Canterbury'. It is famous, however, for its red deer and other game, and Manuka Point Lodge now provides professionally guided trophy hunting.

Ferrying wool from Manuka Point Station across the Rakaia River by a converted bren-gun carrier in the late 1940s.

Courtesy Corinne Crawley

Rangitata and Ashburton Valley sheep stations

Between Mt Hutt and the Rangitata River there are half a dozen pastoral runs occupying the front-country hills, and the two roads through the Ashburton and Rangitata Gorges lead to another 10 stations in the Lake Heron Basin and upper Rangitata Valley.

The hills fronting the Canterbury Plains tend to be relatively wet, with only a small proportion of sunny country. Most of this land was taken up by Acland, Tripp and others in the mid-1850s, but the present properties, apart from **Winterslow**, bear little resemblance to the early runs.

Mount Somers Station, on the east side of the Ashburton Gorge, once covered over 7,000 hectares. However, the mountain is no longer part of the property and little hill or tussock country remains. Most of the lowland has been cultivated and 5,000 deer have been introduced. From 1875 the run was operated by Alfred Peache, who topdressed and cultivated the flat land, and developed a limestone quarry and small coal mine. The station remained in the Peache family until 1983, when it was taken over by Mark Acland, a great-grandson of the original owner.

To the south of the gorge, the hill country that was once part of Mount Posession and the Anama run now supports three stations. Over the past 70 years **Inverary** has been developed by the Chapman family. It has a great deal of oversown and cultivated land, and runs over 10,000 sheep and 1,500 cattle.

Edendale is a small property that proved too damp for sheep; it now runs about 3,000 deer and some cattle. On the north side of the Rangitata is another small run, **Tenahaun**, which operates in conjunction with another property on the plains.

On the road to Lake Heron and Mount Arrowsmith Station.

The road up the Ashburton Gorge from Mount Somers village passes through part of Mount Possession Station on the left. **Barrosa**, to the right, has a fine homestead, built before the property was subdivided from Clent Hills in 1918. The hill country is mainly used for cattle, while further inland, around the Maori Lakes, the flatter, drier country carries sheep.

After about 25 kilometres the road emerges from the gorge and forks at the Hakatere Station homestead. The road to the right gives access to runs on the glacial moraines that make up the floor of the Lake Heron Basin. This area is at high altitude and snow frequently falls in winter, but summer is hot and dry. The small **Castle Ridge** run was part of Barrosa until 1992 and, in spite of the climate, is sufficently well developed to run about 9,000 sheep.

Clent Hills, the centre of the large area taken up by Francis Leach in 1857, was once a substantial run, occupying the flats and hill country to the east of the road. The fortunes of this property fluctuated for the

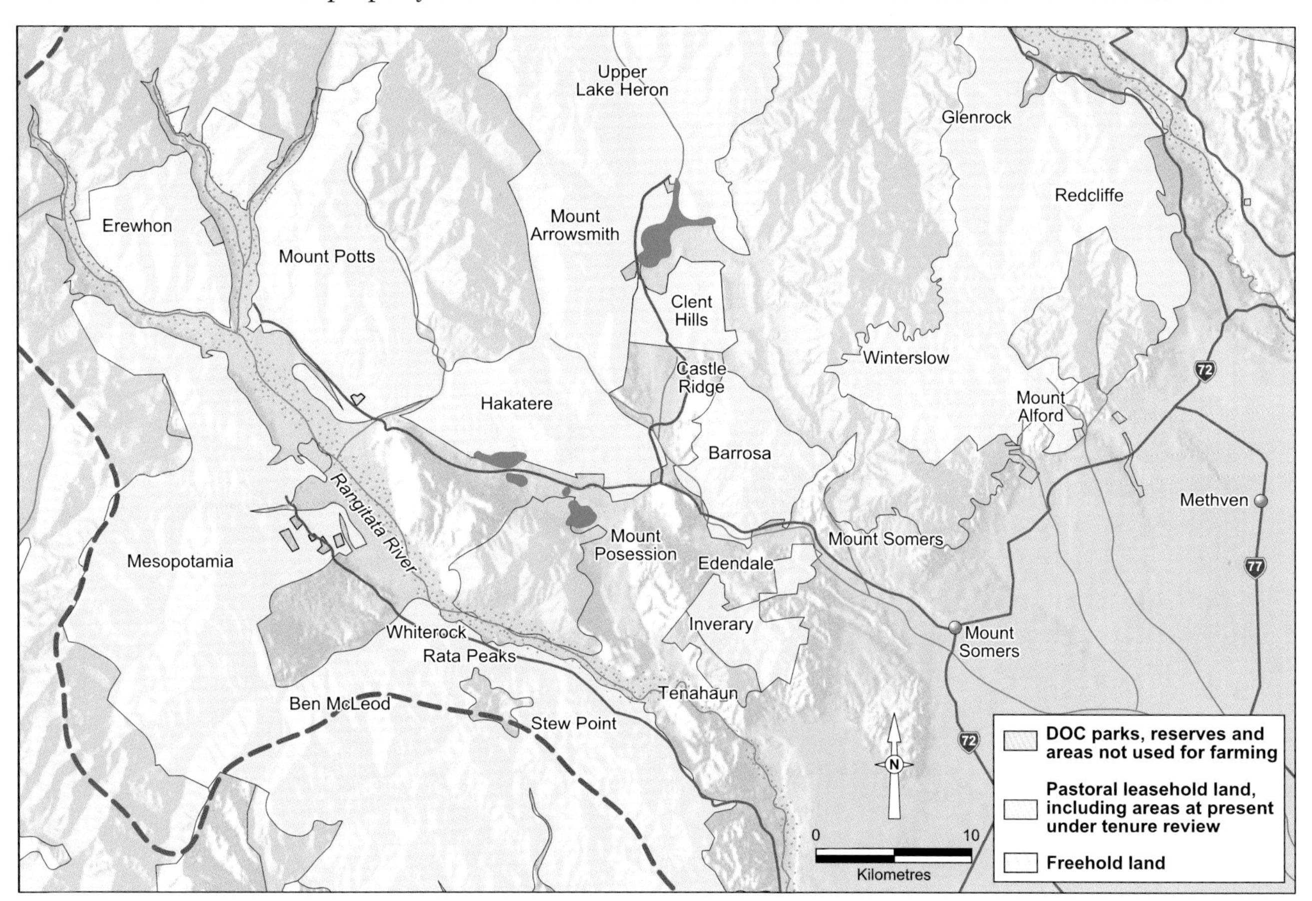

Rangitata and Ashburton Valley sheep runs.

Cartography by Tim Nolan

first 60 years and, when taken over by Bob Buick in 1923, it was plagued with rabbits. Buick, then his son, ran Clent Hills until 1982. Once pests were under control, innovative management developed the property to the stage where it carried 7,000 merino sheep as well as cattle. In 2004 most of the run, including a number of the small lakes, was purchased by the Nature Heritage Fund and gained protection as a conservation park.

The two most remote properties in the area are **Mount Arrowsmith** and **Upper Lake Heron** Stations, both of which remain traditional high-country runs – isolated among rugged mountains and challenging to farm during long winters. The homestead of the former is on the western edge of Lake Heron and the property runs high into the Arrowsmith Ranges. It was home to several generations of the Urquhart family, who ran it from 1926 to 1974, and provides a popular access point for trampers and mountaineers.

Upper Lake Heron, at the end of the road, is also associated with a high-country dynasty. The Todhunters took over this property in 1917 and it is now farmed by a fourth-generation family member.

The left fork of the road at Hakatere leads to three large runs on the north bank of the Rangitata River.

For nearly 20 kilometres the road passes through broad flats bordered by hills that, on the north side, are part of the **Hakatere** run, and on the south side, part of **Mount Possession**. Until they were combined in the 1920s, these runs had changed boundaries several times and were often at the mercy of rabbits, stock loss from heavy snowfalls, and low wool prices. Sam Chaffey was appointed manager in 1924 and quickly earned a reputation as an innovator, introducing pest control and aerial topdressing.

Mount Possession/Hakatere now carries large deer and cattle herds as well as 13,000 sheep on its 27,000 hectares.

The road drops down to river flats on **Mount Potts** Station. This property includes steep hill country behind the homestead and, regarded as uneconomic on its own, has at various times been absorbed into Mount Possession or Erewhon. In the 1960s Arthur Urquhart developed a ski area and accommodation complex on the property. The flat areas were topdressed, irrigated and cultivated, enabling it to operate as a separate run with cattle and deer as well as sheep.

Erewhon (an anagram of 'nowhere' and named after Samuel Butler's novel) is at the end of the road. For a while known as Stronechrubie, this property was taken up in 1861 and its early history was fraught with disasters – crippling snow, river-crossing problems, bankruptcy and, in the 1930s, rabbits and deer. In 1943 a new era was ushered in by Arthur Urquhart, who improved the amount and quality of wool grown

Above: The old Stronechrubie (Erewhon) homestead in winter.
Courtesy Colin Drummond, Erewhon Station

Left: This view of the Erewhon homestead area shows cultivated flats in the foreground with the glacially shaped Jumped Up Downs blocking access to the Clyde Valley behind.

The wide bed of the Clyde River near Erewhon looks negotiable in dry conditions, but after rain it can pose major problems in moving stock to and from the Cloudy Peak block on the west side.

on the property and developed safari hunting and skiing operations. His sons continued to topdress the lower country and built up a deer herd. Erewhon was sold to Colin and Christine Drummond in 1995.

A block of over 40,000 hectares on the south side of the Rangitata as far up as Forest Creek was secured in 1855 by John Acland and Charles Tripp. This became the original **Mount Peel** Station, and the road up the river passes the homestead built by Acland and still occupied by his descendants. Most of the original property has now been subdivided into a smaller runs, including **Stew Point**, **Rata Peaks** and **Whiterock**. **Ben McLeod** Station lies in the valley and hill country to the south.

Forest Creek marks the eastern boundary of **Mesopotamia** Station, whose 28,000 hectares extend to the Main Divide. The success story of the original runholder, Samuel Butler, is told on page 76. After Butler's departure, Mesopotamia slipped into the

The original homestead at Mesopotamia in 1871, a few years after Samuel Butler left the property.
Alexander Turnbull Library MNZ-0386-1/4F

The present Mesopotamia homestead is hidden by trees (which also cover the site of the painting reproduced on page 151). In the background are the high peaks of the Two Thumb Range.

Ed is a big, rowdy huntaway, and Midge is a smart little heading dog. In spite of changing management methods, dogs such as these still play a crucial role in high-country farming.

familiar high-country malaise of frequent changes of ownership, depletion of the flock by heavy snows, several periods of financial problems and invasion by rabbits.

The arrival in 1922 of Ron Urquhart as manager, a role carried on by his son, brought a period of stability and increasing stock numbers until the 1930s Depression and further rabbit plagues checked progress.

Malcolm Prouting took over as manager in 1943, and soon owned the property. Through the 1950s new buildings, roads and bridges were built, aerial topdressing was begun, groynes were constructed to control flooding and, in 1957, a school was built. The station is now operated by Laurie Prouting, a Cessna plane is used for sheep spotting, a good deal of winter feed is grown on the flats, and 18,000 sheep share the grazing with 400 cattle and 1,400 deer. The homestead area has become almost a village, a far cry from Butler's solitary cob cottage.

The sheep stations throughout the central Canterbury high country share similarities in their history and development. Most had uncertain beginnings and weathered hard times before enjoying a long, stable period through the mid-twentieth century, often passing from father to sons. During the 1950s a dozen or so smaller blocks were subdivided from several of the easier runs, particularly in the lower Rakaia Valley, for soldier-settlement farms. Apart from these, there is also a common thread in the runholders' responses to the increasing pressure for change. As much of the more difficult higher country is being retired from grazing, an emphasis is being placed on making the most of the better land. Stock diversification has seen cattle and deer introduced, and tourism ventures are being explored.

Coal and hydro-electricity

Although the primary reason for the settlement of the high country was to carry out pastoral farming, there were several enterprising attempts to develop other forms of economic activity. The most quixotic of these was the unsuccessful attempt to find gold in the headwaters of the Wilberforce River in 1882. Other schemes have been rather more soundly based.

New Zealand's first hydro-electric power station, at Lake Coleridge, began operating in 1914, producing 4,500 kW of electricity. Expansions to the plant boosted the output to an impressive 35,000 kW by 1930. The station is still operating, though its contribution to the national grid is now minuscule.

Between 1900 and 1910 increasing demands for electric power prompted the Christchurch City Council to explore various possibilities for generation. The favoured option for most of this time was a large dam across the Waimakariri River at the mouth of the gorge, and detailed feasibility plans were drawn up. However, these were abandoned following a more favourable report on Lake Coleridge, and work started on the project in 1911. This called for water from the lake to be piped two kilometres through moraine on the south side of the lake to a surge chamber, then drop down penstocks to the power house, 150 metres below, and discharge into the Rakaia River.

Below is a brief summary of the fascinating history and development of the power station, which is recounted in full by Rosemary Britten in her excellent book, *Lake Coleridge: The power, the people, the land.*

Building the Coleridge power station involved extraordinary solutions to overcome problems. The isolation of the area presented a major difficulty: heavy equipment and stores had to be transported 50 kilometres from Coalgate, mostly hauled by as many as 15 traction engines. Workers toiled through frigid winters and were exposed to the full force of nor'westers blasting across a tussock-covered plain. On the lake, waves created by the wind caused problems of shifting shingle, which regularly spilled into the intake, and building the tunnel through moraine gravels and clays was a unique challenge.

Early days at the Lake Coleridge power station site. The second stage of construction is not complete and the main plantations have not yet been established.
James Brake Collection, Canterbury Museum 1968.213.111

Looking down the penstocks to the Lake Coleridge power house, with the Rakaia River in the background.

As many as 400 men were employed at times on the construction project, many of them living in tents on the top of the hill while working on the intake and tunnel, and others on the terrace below, working on the penstocks and power house. The first electricity was generated in 1914 and demand grew so fast that within 10 years the scheme was enlarged. Major work was necessary at the head of the lake to increase the water supply. First, a permanent diversion of the Harper River into the lake was excavated and, more recently, a canal was built to draw water from the Wilberforce.

The village housing permanent staff grew and became a more comfortable place to live as the many trees planted began to provide shelter. Although contact with the outside world was provided by a bus running from Coalgate three times a week, the settlement remained a close-knit community with about 30 houses, a hostel and huts for single men, a social hall, a small primary school and a post office.

In recent years, with improved access and increased automation of the power house, staff numbers have been greatly reduced and at least half of the houses in the village have been made available for sale to the general public. In 1995 the school was closed and pupils now travel to Windwhistle, 20 kilometres away. Although the grounds and trees around the village and power house are still well kept and the former hostel is now run as a tourist lodge, the settlement has a rather lonely feeling, as if left behind by the world.

Early in the development of the region three separate ventures were set up to mine the coal seams in Tertiary strata exposed on some of the lower hill country. These were derived from beds of peat laid down as long as 60 million years ago, then compressed and uplifted in the Kaikoura Orogeny. The result was a series of small, rather faulted beds of mainly peaty brown lignite.

The Mt Somers mines operated intermittently for over a hundred years in the hills on the north side of the Ashburton Gorge, with the first licence applied for in 1867 by Charles Tripp. About 10 small-scale open-cast mines were established in the area, each eventually failing because of the broken nature of the seams or from fire damage. The last closed in 1968. Most of these diggings were on the ridge running alongside Woolshed Creek, where the remains of the Blackburn mine, along with a tramline used to transport the coal, have survived.

The lower Broken River area, up the Avoca Stream in the Waimakariri Valley, was the location of another mining endeavour. In 1865 John Enys in his diary noted the presence of '5 large coal seams within 10 miles of Castle Hill'. A lease was taken out on this resource in 1915, and three years later Mount Torlesse Colleries sank an underground mine. During

The remains of the Blackburn coal mine on Mt Somers.

the next nine years as much as 72,500 tonnes of high-grade lignite coal was taken out and, at the peak of its operations, the company employed 58 people. The coal was transported by wagons on tramlines five kilometres to the Midland Railway line at Avoca. A serious fire in the mine in 1924 was followed by a series of lesser outbreaks and these, together with diminishing returns, resulted in the closure of the operation in 1927.

The smallest of the mines was on the west side of the Acheron Stream, on the north side of the Rakaia, where anthracite – a high-grade, hot-burning coal – had been converted from lignite by volcanic activity. This was briefly mined at the beginning of the twentieth century, but the coal was found to burn with too much heat for general use. The Stuart family re-opened the mine in the mid-1930s and extracted 5,000 tonnes of coal between 1941 and 1945. The seams were up to three metres thick but were heavily faulted and difficult to work. A severe flood in 1951 destroyed much of the infrastructure. New shafts were worked until the 1960s, and briefly in 1982, but though the coal was regarded as being of good quality, it proved uneconomic to mine.

The Avoca mine in the Broken River area in its heyday.
W. A. Kennedy Collection, Canterbury Museum 1975.203.44

CHAPTER EIGHT

Alpine Playground

Traditional recreation

The central Canterbury high country has been enjoyed for recreation since the early days of settlement. The explorers carried guns to shoot birds for the pot, and sports hunting, fishing and tramping were all reported from the 1880s. Skiing was introduced in the late 1920s, but the earliest recorded organised recreational activity was ice skating.

Ice skating

Lady Barker's description of skating on Lake Ida (see opposite) indicates that, as early as the 1860s, the wealthier settlers at least were looking for recreation in the high country. In 1868 John Enys of Castle Hill Station recorded in his diary that he went skating on Lake Pearson, and an 1870 entry contained a laconic lament: 'Skating on Lake Lyndon; got under the ice.' Skating is possible on several of the smaller lakes dotted through the high-country valleys and basins, and several homesteads have ponds that freeze over in winter. Lake Ida is the only significant body of water that is sufficiently shaded to retain skatable ice for long periods. Located in an isolated valley between high hills on Ryton Station, it is reached by travelling a couple of kilometres up a side road off the Harper road. An ice-skating club was based here for many years, and a hut and canteen were built on the site. Although suitable ice conditions occur less frequently nowadays, reports of a good freeze still encourage skating enthusiasts to head up to the high country.

Tramping and climbing

The majority of people in Victorian times regarded the notion of venturing into mountains with apprehension rather than appreciation. As the writer of the Department of Conservation's *Story of Arthur's Pass National Park* neatly put it: 'No one in the early days contemplated heading off into the mountains for adventure or inspiration, except for the occasional character with a glint in his eye and a copy of romantic poetry in his haversack.'

Left: Ice skating on Lake Ida in the 1940s. *Right:* Lake Ida in summer.

There are few reports of these occasional characters tramping or climbing for pleasure in the high country of central Canterbury. The first recorded ascent of a peak in this area was in 1849, by the surveyor Charles Torlesse, who with a Maori companion climbed the highest peak of what is now the Torlesse Range. There is also a record of Mt Hutt being climbed in 1889. Two years later, Mt Philistine and the lower peak of Mt Rolleston, on the southwest side of Arthur's Pass, were claimed by a party that included A. P. Harper, George Mannering and Marmaduke Dixon, each of whom was noted for attempts on summits such as Aoraki Mt Cook. But these climbs were exceptions and it was not until the 1912–13 season that a small group of enthusiasts made first ascents of the major peaks at the head of the Waimakariri – Davie (2,294 metres), Murchison (2,400) and Wakeman (2,286).

George Mannering and Marmaduke Dixon, pioneer mountaineers and adventurers in the Waimakariri area.

E. Wheeler & Son photograph, Canterbury Museum 1971.248.97

Lady Barker skates at Lake Ida

Lady Barker devoted a chapter to ice skating at Lake Ida in her second book, *Station Amusements in New Zealand*. In the mid-1860s she and her husband, Frederick Broome, always keen for fun and excitement, were invited to join an ice-skating expedition. They travelled on horseback to Lake Coleridge Station and then on to the hut of J. C. Monck, a young bachelor struggling to make his fortune by grazing sheep.

> I can never forget that first glimpse of Lake Ida. In the cleft of a huge, gaunt, bare hill, divided as if by a giant hand, lay a large black sheet of ice. No ray of sunshine ever struck it from autumn until spring, and it seemed impossible to imagine our venturing to skate merrily in such a sombre looking spot . . . I despair of making my readers see the scene as I saw it, or of conveying any idea of the intense, the appalling loneliness of the spot. It really seemed to me as if our voices and laughter, so far from breaking the deep eternal silence, only brought it out into stronger relief. On either hand rose up, sheer from the water's edge, a great barren, shingly mountain; before us loomed a dark pine forest whose black shadows crept up until they merged in the deep crevasses of the Snowy Range . . . we tried hard to be gay, and no one but myself would acknowledge that we found the lonely grandeur of our 'rink' too much for us. We skated away perseveringly until we were both tired and hungry, when we returned to Mr K**s hut, took a hasty meal, and mounted our chilled steeds.

The road over Arthur's Pass provided access to the interior for small numbers of outdoors enthusiasts and, by the turn of the century, guides from the Glacier Hotel at Bealey Corner were taking groups up the Waimakariri to see the small glaciers at its head. It was the completion of the Midland railway in 1924, however, that really opened up the area to visitors. Five-shilling rail excursions to Arthur's Pass resulted in an influx of weekend trippers from Christchurch, eager to enjoy the dramatic scenery and make the most of tramping, climbing and skiing opportunities. The first Carrington Hut, four to six hours up the Waimakariri at its junction with the White River, was built by a pioneering tramping and mountaineering club in 1929 to provide accommodation for those venturing into the Main Divide. The explosion of interest in the mountains during the 1920s and 1930s was described by John Pascoe, a renowned climber and writer, as 'the Waimakariri invasion'. In the first Arthur's Pass National Park handbook, published in 1935, Pascoe described no fewer than 120 separate routes up 35 peaks in the area; he himself climbed all these over a period of 20 years.

Access into the heart of the mountains, the close proximity to spectacular scenery, and relatively easy routes up each of the main tributaries of the Waimakariri ensured that this area became a second home for many trampers and mountaineers. The Friday night ritual of riding in the guard's van of the 'Perishable' goods train and spending a strenuous weekend in the mountains was the beginning of a love affair with the great outdoors for several generations of young adventurers from Christchurch.

The rise of interest in the upper Waimakariri high country in the 1930s and '40s stimulated the development of tramping tracks and routes over alpine passes and the building of mountain huts and bivouacs, especially by the Canterbury Mountaineering Club and the New Zealand Alpine Club. In the following two decades more huts were built, mainly by the Forest Service, for the use of hunters attempting to control deer, chamois and possums. The laborious task of backpacking loads of building materials was eased from the end of the 1960s by the use of helicopters. There are now more than 30 huts and bivouacs on the Canterbury side of the Arthur's Pass National Park and another half-dozen in the surrounding Waimakariri valleys.

Triangle Hut, in the Avoca Valley, pictured here in 1952, was typical of many built about this time.

The northeastern side of the Arthur's Pass region is particularly good tramping country, with attractive routes up beech-forested valleys such as the Poulter, Hawdon and Edwards leading to good huts and to passes above the bushline. On the southwestern side, the Waimakariri headwaters tend to have somewhat more challenging routes in higher and more

Building a base

The construction of the first Carrington Hut involved enthusiasm, determination and misadventure. The young climbing devotees who formed the Christchurch Tramping Club in 1924 had as one of their first goals the building of a hut up the Waimakariri River to provide a base for expeditions into its headwaters and surrounding peaks. This project began with the club secretary, Gerard Carrington, taking packhorses up the valley with loads of timber and corrugated iron, which were stored at various places along the track. Unfortunately, Carrington was drowned the next year while rafting through the Waimakariri Gorge, and with him went the cache-location map. Other members continued the work of fundraising and carrying materials. Once, a packhorse bolted, scattering timber over the tussock flats, and one member claimed that he had toiled 800 kilometres up and down the Waimakariri riverbed in 1928–29.

A man named Cox was given a contract to complete construction, but this was further delayed when he was declared bankrupt. Although the original building was cold and smoky, uncomfortable and dark, it was always a welcome sight to those trudging up from the road late at night or returning from a day's strenuous climbing. The present Carrington Hut, dating from 1974, is much more comfortable and well equipped with 36 bunks.

Photo of the first Carrington Hut from Logan, *Waimakariri*

Climbing in the Garden of Eden, a high plateau at the head of the Clyde River.
Nick Groves

rugged mountains. A classic tramp is the Three Pass Trip, which leaves from Carrington Hut and crosses the Main Divide at the Harman Pass, returns over the Whitehorn Pass to the headwaters of the Wilberforce River, then uses the Browning Pass to access the Arahura Valley and the West Coast. Many trampers have tested themselves on this route and have often found the toughest challenge to be the fickle and sometimes hazardous alpine weather conditions.

The mountains to the southwest of Carrington Hut are popular as a training ground for aspiring climbers, who can refine their skills in the Shaler Range at the head of the White River. For those looking for easier climbing, the Craigieburn Range is readily reached from the skifield roads.

The main ranges of the Southern Alps in the headwaters of the Rakaia and Rangitata Rivers also provide considerable challenges for mountaineers. However, they all involve long bashes up riverbeds, and have never become as popular as peaks in the Arthur's Pass area.

At the head of the Wilberforce, Mathias and Rakaia Rivers are significant summits such as Mt Whitcombe (2,644 metres). Further south, the Cameron Valley west of Lake Heron leads to the jagged range named for Mt Arrowsmith (2,798 metres), which tests the most experienced of climbers. The headwaters of the Rangitata includes an ice plateau, known as the Garden of Eden, at the head of the Clyde River; and Mt D'Archiac, up the Havelock River, is the highest in our region, at 2,875 metres.

There are a number of huts up these valleys – half a dozen at the head of the Rakaia alone, others up the Mathias, Wilberforce and Avoca, and in the tributaries of the Rangitata, Lawrence, Clyde and Havelock Rivers. Many of these huts were erected for deer cullers or musterers and are often small and, in some cases, in poor condition.

The impressive peak of Mt D'Archiac, at 2,875 metres an enticing challenge for serious mountaineers.
Courtesy 4 x 4 New Zealand

Trampers on Mt Somers.
Courtesy Tussock & Beech Ecotours

Those looking for closer-to-hand and less technically demanding tramping can choose from a number of tracks in the Canterbury foothills, many of which are suitable for one- or two-day trips. Perhaps the best known are up or around Mt Somers. This area is popular with school groups and also professional guides taking inexperienced tourists on adventure or ecotourism walks.

Hunting

Shooting game was a popular pastime among the upper classes in rural England during Victorian times, and the early Canterbury settlers brought that interest to the colony with them. In 1857, when exploring in the upper Rakaia Valley, Thomas Potts recorded in his diary that his friend Harry shot several ducks (see page 75), and Lady Barker devotes a chapter to pig shooting in *Station Amusements in New Zealand*.

Because there were no game mammals in New Zealand, the early settlers quickly set about filling this void by setting up acclimatisation societies to import suitable species of game mammals, fish and birds. This they did with much enthusiasm and with total disregard for the indigenous species already present, for which we have been paying the price ever since.

The main game animals in our region are red deer, first released by the Canterbury Acclimatisation Society into the upper Rakaia Valley in 1901 and 1902, and in the Poulter Valley in the Waimakariri a few years later (see page 68). The members of the acclimatisation society tended to regard these herds as their own and the beech forests as the rightful habitat of deer. The animals thrived and provided recreational hunting for a few years, but by the 1920s it was generally recognised that this was insufficient to keep numbers down and that deer had become a major pest, particularly in the Rakaia and Rangitata Valleys. Where present in large numbers, they ate out the undergrowth in beech forests and, over time, their selective browsing depleted more palatable plants.

Himalayan tahr and chamois had also been released around Mt Cook and gradually spread into the headwaters of the Rangitata and Rakaia, where they also began to cause serious problems by eating the fragile alpine vegetation.

Following a 'Deer Menace Conference' in Christchurch in 1930, the government established a programme aimed at complete eradication. The men employed as deer cullers were a far cry from the well-to-do recreational shooters of Victorian times. They were generally tough, resourceful and prepared to spend months living rough, often alone or in pairs, in the isolated mountains and bush in the headwaters of the three great Canterbury rivers.

The highly skilled cullers often recorded huge tallies – Bill Chisholm (later manager of Molesworth) and two others shot 3,000 deer and 1,000 chamois on Erewhon Station in three months. In spite of the eradication policy, deer numbers remained high, and the Forest Service took over deer control in the mid-1950s. After much controversy among recreational hunters and conservationists, the eradication policy was abandoned and a combination of recreational and commercial hunting was allowed.

Many fine trophy heads have been produced by the 'red stags of Rakaia', but commercial hunting has been more important in reducing deer numbers, with helicopters enabling much easier access and then facilitating the removal of carcasses. Fortunately, the Canterbury high country has not seen the daredevil antics and bitter wars carried out among the 'chopper boys' in Fiordland.

A new phase in the deer story began in the 1970s with the realisation that there were good commercial possibilities overseas for the sale of venison and other deer products, particularly antler velvet, and several runholders began to farm deer. To the forefront of this experiment was the Urquhart family of Erewhon Station, who had begun offering guided safari hunting of deer, chamois and tahr. Under pressure from recreational hunters, permits for this venture were withdrawn, so in 1972 the Urquharts set up a deer farm. This was very much a trial-and-error operation. The first problem to be overcome was the capture of sufficient numbers of wild animals, achieved initially by surrounding deer grazing on the flats with high netting fences. Later, helicopters were used on Mesopotamia Station to herd wild deer into pens.

In the early 1980s Mark Acland of Mount Somers helped to pioneer capture by helicopter, using various methods ranging from 'bulldogging' (men simply leaping on the back of a deer) to immobilising darts and nets fired from special guns. The industry has developed to the point that at least a third of all the pastoral runs in the Canterbury high country have established deer herds, with Mount Hutt specialising in this and running 15,000 head.

Yet another episode in the story of high-country deer has opened with the rise of guided game hunting and the development of safari parks, aimed largely at overseas clients. This is discussed in the next chapter.

Chamois and tahr (see pages 68) slowly spread north along the high ranges of the Southern Alps from the Mt Cook region, where they had been released, and by the 1930s had reached the headwaters of each of the Canterbury high-country rivers. Both animals occupy the fragile habitat above 1,300 metres, and chamois in particular had become sufficiently common to be included in the deer-eradication scheme. The numbers of chamois and tahr are now considerably reduced and they are much prized by trophy hunters.

Feral pigs also spread into the high country and, although never present in large numbers, became quite common around some bush edges. They provided occasional sport for hunters for much of the twentieth century.

Fishing

On a hot summer's day in 1877 William Izard, a friend of the Enys brothers, walked the 32 kilometres from Springfield to Castle Hill. As he forded the Thomas, the last creek before Trelissick homestead, he stopped to drink from and bathe in a deep pool, where he also tickled three fine trout he found swimming there. Full of pride, he carried them up to the house, only to be told by John Enys, 'I put those in the creek this morning!'

The descriptions of trout releases in Enys's diary are among the first records of trout being liberated into the Canterbury high-country streams following the acclimatisation society's enthusiastically embracing the idea of stocking local rivers with imported fish.

Red deer stags on the High Peak Game Estate.

A helicopter retrieving a tahr shot on an alpine ridge in the upper Rakaia. Chamois also thrive in this rugged environment.
Courtesy Back Country New Zealand

Neither the tiny native fish nor the abundant eels, favoured as food by Maori, were suitable for sports fishing, so in 1867 the society brought 800 brown trout eggs across from Tasmania. It took some years to establish successful populations, but by the mid-1870s these fish had been introduced to most Canterbury rivers and lakes, often by runholders who were keen anglers. In 1868, for example, John Oakden of Acheron Bank released 20 trout into Lake Coleridge. At the time no thought was given to the fate of the variety of the small native species, which were generally unable to compete with aggressive newcomers.

Brown trout found an excellent habitat in the headwaters of the Canterbury rivers, where they found abundant food supplies, good gravel spawning grounds, stable banks in the smaller streams, clear water and a lack of disease or predators. They are now present in most rivers in the region, growing to 1.5–2 kilograms, and are sufficiently common for there to be no need for stocks to be replenished from hatcheries.

Rainbow trout, released in the 1880s from California, are also present in some high-country streams but are more common in the small lakes, a habitat they prefer in this region.

Several other North American trout species have been introduced into Canterbury lakes, where tiny populations have survived. Mackinaw (lake trout) have been present for over a hundred years in Lake Pearson, and brook char in Lake Emily since 1939.

Between 1864 and 1910 there were more than 20 attempts to introduce Atlantic salmon into South Island rivers, all of which failed. However, the big braided Canterbury rivers offered ideal conditions to quinnat salmon. These fish spawn in the upper reaches of the Waimakariri, Rangitata and especially the Rakaia. Several months after hatching, the fingerlings make their way downstream to the sea. After two to four years adult salmon return and migrate upstream to their home waters to spawn and then die.

The Rakaia consistently has the largest salmon runs, averaging around 10,000 fish each season. Runs in the Waimakariri and Rangitata are a little more than half this total. The numbers vary greatly from season to season, for reasons that are constantly debated, and may be assisted by the release of young fish from a hatchery near the mouth of the Glenariffe and Double Hill Streams, which are favoured spawning areas in the upper Rakaia.

Most salmon fishing takes place at the mouths of the big rivers, but there is also good sport at times further upstream. The upper reaches of the Waimakariri are closed during the spawning season.

This stable stream near Lake Coleridge provides an ideal spawning ground for trout.
Courtesy Back Country New Zealand

While not as celebrated as some other fishing areas of New Zealand, the headwater streams and scattering of glacial lakes of the central Canterbury high country have always been regarded by those in the know as something of an anglers' paradise. The combined attractions of grand scenery, uncrowded waters and good-sized, fighting brown and rainbow trout attract devotees year after year.

Even a fishing paradise has its drawbacks, however, and in the high country a persistent inconvenience is the nor'wester. Fishing guide books to the region frequently warn: 'Exposed to the nor'westers', 'Good fishing if there is no nor'wester' or 'Fish hard to spot in the nor'wester'. Considerable skill and forbearance is required to cast into a strong wind with matagouri bushes at your back.

The following are some of the region's favoured fishing areas, given suitable conditions.

The Waimakariri Gorge is most easily accessed by jetboat, and good brown trout can be found in the more stable upper tributaries such as the Poulter River. In the Cass area, the small glacial lakes – Grasmere, Sarah and Hawdon – carry both brown and rainbow trout, and the larger Lake Pearson has a colony of tiny anglers' huts scattered among the willows on its shoreline.

On a calm day the shores of Lake Clearwater provide good spotting opprtunities for anglers.

Fishing high-country style

I spent my childhood on a property that included the headwaters of the Selwyn River, a fine brown trout stream, where my mother developed the skilful but illegal art of 'tickling' – gently feeling under a stream bank for a lurking trout, then flipping it out of the water.

On one occasion a visiting angler, unaware of my mother's prowess, tried to pursuade her to try fly fishing. She initially declined but, under pressure, took a rod and was shown how to cast. She headed upstream while the angler tried his luck downstream. When she returned with a fine fish an hour later, he was empty-handed and full of praise for her first-time effort. He was puzzled, though, by the fact that the fish seemed to have been hooked in its lower jaw, something he had not seen before. My mother did not enlighten him about her technique.

Some time later, word of my mother's poaching skill reached the local policeman, who took the opportunity at a social occasion to give her a little warning. She went straight home, tickled a very nice trout and sent it to the policeman via the local bus the next day. That evening he rang to thank her for the venison.

After 60-odd years, perhaps this story may now be safely told.

Further south, Lake Coleridge contains not only rainbow and brown trout but also landlocked quinnat salmon. This large body of water is very exposed to nor'west and southerly storms and has limited shoreline access. Nevertheless, it is very popular with anglers, and during the season many camp on the flat at the mouth of the Ryton River, where there is shoreline access, boats can be launched and shelter found for campervans and tents in clearings surrounded by matagouri scrub. Further along the Harper road the occasional angler camps beside attractive little lakes such as Georgina and Evelyn. Lake Selfe is generally regarded as having the best fishing, producing brown trout up to three kilograms.

The road up the Ashburton Gorge leads to another group of glacial lakes – Heron, Clearwater, Camp, Emily and several others – that have rainbow and especially brown trout populations. These waters are all rather exposed to the wind, but the cluster of baches on the shore of Lake Clearwater indicates that some anglers are not detered by the likelihood of adverse weather. Lake Heron, which drains into the Rakaia River, also boasts a population of good-sized quinnat salmon.

The braided strands of the Rakaia and Rangitata Rivers are too silty, unstable and prone to floods to provide good trout habitat, but the annual run of salmon does attract anglers.

Skiing

Although emerging relatively late as a sport in the Canterbury high country, skiing (and its recent sibling, snowboarding) is by far the region's most popular recreational activity. Every winter and into spring many thousands of people, local and from overseas, visit the eight fields that have been developed to varying degrees.

Cantabrians' interest in skiing was initially focused on the area around Arthur's Pass after Guy Butler set up a hostel in the village in 1927 and offered eight pairs of Norwegian skis – the entire stock in Christchurch and Dunedin – for hire. Within a day of the hostel's opening, lady guests 'in long black skirts and cloche hats, were sliding cautiously over the snow'.

The second stimulus was the arrival of Oscar Coberger, a German immigrant who had been a guide and ski instructor at the Hermitage, Mount Cook. In 1929 he imported a stock of skis and set up a shop at Arthur's Pass. In the same year the Christchurch Ski Club was launched, and from 1930 the Railway Department's cheap Sunday excursions greatly boosted the popularity of skiing, with some weekends attracting over a thousand trippers.

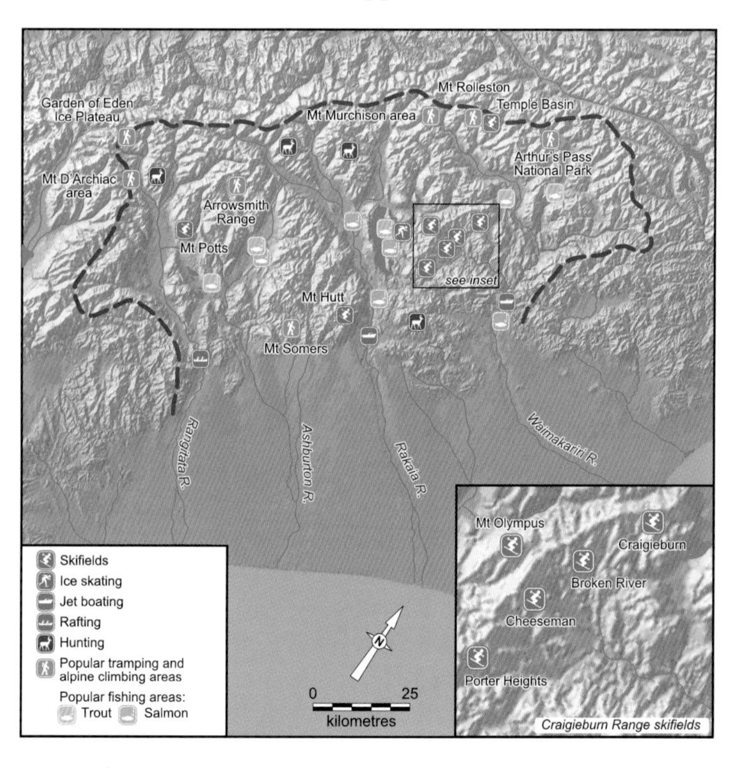

The main recreational areas in the central Canterbury high country.
Cartography by Tim Nolan

Skiing on roadside slopes at Arthur's Pass in the late 1920s.
Photo from Logan, *Waimakariri*

Temple Basin

Enthusiasts soon began to search for better areas to ski than the roadside slopes near the top of the pass, and by 1933 the Temple Basin field had been established and a hut built a steep 500-metre climb from the road. This skifield is small and high (about 1,600 metres), the weather is often bad because of its proximity to the Main Divide, and most slopes are more suitable for experienced skiers. It now has tows and accommodation for 120 people, but retains a casual atmosphere and has survived in the face of competition from the larger and more accessible fields that have developed in Canterbury.

Club fields in the Craigieburn Ranges

These small venues – established in 1929 and from the 1940s – had the attraction of being closer to Christchurch by road than other alpine areas. Each of the four fields was planned and developed by club members who put in endless hours building access roads, huts and ski tows – with limited resources but a great deal of passion. These fields also now cater for visitors and have a well-deserved reputation for providing a friendly, easy-going environment and good sport on uncrowded slopes.

At about the same time as Temple Basin was being opened up, investigations began to develop a field on the slopes of Mt Cockayne in the Craigieburn Range. In 1929 the Cheeseman field was set up by the Canterbury Winter Sports Club and, in a very busy first year, members fashioned a six-kilometre track, built a hut at the road end, a toboggan run and an ice rink. After the war a proper road was put through, and by

1950 the facilities included a tow and a hut on the skifield. Thirty years later, the Cheeseman club had a thousand members and a waiting list.

The field is quite high – 1,845 metres at the top and with a vertical run of 293 metres. Popular as a family venue because of relatively easy skiing, it is now reached by 12 kilometres of unsealed road through beech forest, and there is accommodation both at the base in the Forest Lodge and on the field at the Snowline Mountain Lodge.

Craigieburn field was established in 1947 from a base six kilometres in from the main road by a ski club in association with the Canterbury Mountaineering Club. It now has several lodges, three rope tows and a membership of about 350. Promoted as 'steep, deep and cheap', it is particularly suitable for experienced skiers.

In the late 1940s a basin between the Craigieburn and Cheeseman fields was developed as the Broken River field by the North Canterbury Ski Club, based in Rangiora. Its facilities were built 15 minutes' walk in from the end of a gravel road through the beech forest. Today, three lodges serve the club's 400-strong membership plus visiting skiers. The sheltered slopes are served by five rope tows and the field is usually open until late in the season.

Porter Heights, the most recently constructed field in the vicinity, was opened at the southern end of the Craigieburn Range in 1967 as a commercial rather than a club enterprise. Its access road climbs about six kilometres up the Porter Valley to a plateau, where there is a car park, lodge and cafeteria. The closest skifield to Christchurch, it is also the best equipped of the Craigieburn venues, with snow-making and grooming facilities, and five lifts including a T-bar with a vertical lift of of 750 metres.

Private fields

Mount Olympus, a small field on Ryton Station near Lake Coleridge, was established in the 1960s by a group who formed the Windwhistle Ski Club. They have built a hut and rope tows, and although Olympus is not as accessible as most of the other fields, regular users enjoy its friendly ambience.

Another very small field has been operating for over 30 years on the Mount Potts Station in the upper Rangitata Valley. This is the highest skifield in the South Island, access is difficult and facilities are few. A lodge accommodates 14 guests, and the owners have recently introduced helicopter and snowcat transport to the top of each run.

Mount Hutt

This field is considerably larger than any other Canterbury operation. It is perched at an altitude of about 1,600 metres, and each year many thousands of people make the journey up the tortuous access road to enjoy sport on the slopes. On some days there are more than 4,000 visitors, and in the 2004 ski season about 100,000 skier days were recorded – far more people than visit the rest of the region's facilities.

As many as 300 people are employed on the slopes, while Methven, the nearest town, benefits by more than $12 million every year. Yet Mount Hutt was a late arrival on the skiing scene.

In 1969, nearly 40 years after the first Canterbury fields opened, the possibility of establishing one high in Pudding Hill Basin on the eastern slopes of the mountain was first seriously considered and the idea taken up by enthusiasts from Methven.

The first and greatest problem was to obtain access. In 1972, against all predictions, a pilot road was completed in less than three months by Ashburton

Looking down the Mount Hutt skifield to the Canterbury Plains.
Miles Holden, nzski.com

Carving up powder snow on the Mount Hutt skifield.
Miles Holden, nzski.com

The ancient sport of curling, also called 'the roaring game' because of the sound of the granite stones on ice, was brought to New Zealand by Scottish immigrants. This 'bonspiel' (curling competition) is taking place on a frozen pond near Mount Somers.
Courtesy Tussock & Beech Ecotours

engineering contractors Doug and Keith Hood. This was a remarkable achievement and the trials faced by intrepid bulldozer drivers while pushing through the 12-kilometre road make scary reading. One machine slid 50 metres down a sleep slope, and another was buried in a rock slide. Near the top it was necessary to blast a way through rocky bluffs.

Once the road was completed, it was quickly realised that what had begun as a small community project had the potential to become a major commercial venture, and so the Mount Hutt Ski Field Development Company was set up. Willi Huber, an Austrian immigrant and very experienced mountaineer, spent the winter of 1972 in a small hut at 2,000 metres to investigate the winter conditions on the proposed ski slopes. He built a 250-metre-long rope tow and, for the next eight years, was manager of the skifield. On the opening day in 1973 a staff of 10 and 500 skiers were present.

Interest in Mount Hutt grew rapidly, and by the end of the first season it was already rated as one of the top three ski fields in New Zealand. During the 1970s facilities developed rapidly. Overseas skiers, especially Australians, visited in large numbers, and the township of Methven prospered. A T-bar lift was installed, a heli-skiing operation was set up in 1979 and snowboarding began in the mid-1980s and rapidly gained young converts attracted by its radical image.

By 1986 pressure for further development led to a comprehensive upgrading programme being undertaken. As a result, Mount Hutt successfully hosted the 1990 Alpine Skiing World Cup. Following several difficult years, the field was taken over in 1994 by the Mount Cook Group and, with innovative management, its popularity increased even more. It is now run jointly with Coronet Peak and the Remarkables as a stand-alone company.

After 30 years the Mount Hutt field is now a very successful operation, with quad and triple chairlifts, three T-bars and several other tows, and provides skiing, snowboarding and heli-skiing for thousands of enthusiasts, as well as being by far the largest employer of any business enterprise in the central Canterbury high country.

CHAPTER NINE

Jetboats to Helicopters

Tourism comes to the high country

A family from the city is taken by a four-wheel-drive vehicle to a 1,600-metre peak to watch the sun go down while a barbecue is prepared; an American trophy hunter shoots a tahr on an alpine ridge where he and his guide have been ferried by helicopter; a group of mountain bikers traverse a scree slope high on the Craigieburn Range . . .

These are just a few of the wide range of experiences enjoyed by visitors to the central Canterbury high country. The traditional pursuits of tramping, mountaineering, shooting, fishing and skiing, described in the previous chapter, still attract devotees. Here we concentrate upon a quiet revolution that has

New high-country tourism activities

ecotourism experiences	jetboating
four-wheel-drive tours	mountain biking
garden tours	mountain running
golf	multisport racing
guided fishing	rafting
helicopter flights	rock climbing
heli-skiing	safari hunting
hot-air ballooning	walking
horse riding	windsurfing

Kayaking in the Waimakariri River.
Paul Farrow

www.backcountry.co.nz
backcountry
NEW ZEALAND
BIGHORN
BGB358
GT RADIAL

seen the region embrace growing numbers of adventure-hungry New Zealand holiday-makers and overseas tourists.

Many high-country runholders are now looking to tourism to supplement their income and are offering homestay or luxury-lodge accommodation and access to various forms of recreational activity. At the same time many enterprising operators based in Christchurch city and inland towns such as Methven and Geraldine are setting up ventures aimed at a diverse range of visitors – from young thrill-seekers to elderly folk wanting a day in the mountains, and from backpackers to overseas tourists willing to pay handsomely for extraordinary experiences.

The range of activities on offer is based around the natural assets offered by the high country – spectacular scenery, an unspoilt, uncrowded environment – combined with high-tech means of reaching and enjoying these attractions.

Of the 45 or so sheep runs within our region, about a third offer some kind of visitor accommodation and nearly a half provide one or more of the activities listed in the panel on page 129. Most of these enterprises have been set up within the last 20 years and, apart from two or three, operate on quite a small scale, but collectively they represent a strong trend towards creating a recreational playground in the high country.

A browse through the websites that advertise recreational activities and accommodation in the region reveals a wide range of options. This chapter summarises those available in the three high-country valleys, then looks at some examples in more detail.

The Rakaia Valley

The small group of properties on the north bank of the Rakaia River has been to the forefront in recent tourism developments in the region. The dozen or so operators involved have co-ordinated their advertising and work together under the umbrella of the Lake Coleridge Basin Tourism Group.

About half the members of this association offer accommodation, ranging from large holiday homes at the Terrace Downs Golf Resort (see page 133) to campsites at the Rakaia Gorge and the mouth of the Ryton River. Glenthorne Station and Middle Rock have self-catering lodges, and those at Lake Coleridge power station village and Ryton Station (page 132) can also provide dinner, bed and breakfast.

Most properties also provide guided farm tours and can organise salmon or trout fishing and jetboating trips. Middle Rock and Peak Hill Stations offer garden tours.

The runs to the south of the Rakaia River have not yet embraced tourism to the same extent as those on the north bank. A notable exception was Mount Hutt Station, which converted the impressive homestead, dating from the 1880s, into an accommodation and recreation centre. Unfortunately, this has recently closed down, but accommodation in this area can be found near the Rakaia Gorge bridge, which is also the base for local jetboat trips.

Methven, a growing tourist town about nine kilometres out on the plains, developed largely to service the Mount Hutt skifield, but it is also the headquarters for several businesses specialising in safari hunting, guided fishing, hot-air ballooning and heli-skiing.

The Waimakariri Valley

Recreation and tourism have been a feature of the Waimakariri Valley for much longer than in the Rakaia, largely because of good road access and the creation of Arthur's Pass National Park and Craigieburn Forest Park. Several villages of holiday homes have grown up in the area, first at Arthur's Pass and, more recently, on the sunnier spur opposite the Bealey Hotel. Castle Hill village includes a lodge belonging to St Andrew's College that is used as an educational facility.

Opposite

Snowboarding on Mount Hutt skifield.
Miles Holden, nz.ski.com

Mountain biking in the Poulter area.
Courtesy Dave Mitchell

Four-wheel-drive vehicle taking tourists across a back-country river.
Courtesy Back Country New Zealand

Looking down on Wilderness Lodge Arthur's Pass, nestled in beech forest near Bealey.

Flock Hill was the first Waimakariri sheep station to embrace the tourism industry. It now has a complex offering motel-style accommodation, meals and conference facilities. Accommodation is also available at Castle Hill village, along with the services of a local fishing and skiing guide.

The more recent trends in tourism are typified by the developments at Grasmere and Cora Lynn. The former has homestay accommodation at the upper end of the market and has recently prepared freehold building sites for sale.

Wilderness Lodge Arthur's Pass is located in the lower beech forest on the old Cora Lynn sheep run and has 20 bedrooms and a restaurant. The ecotourism activities it promotes are described on page 137.

The distinctive limestone rock formations at Castle Hill are a particular focus for recreational activities, described further on page 139.

The Ashburton and Rangitata River Valleys

High-country tourism has been slower to penetrate this region, though the foothills area has a well-organised and extensive range of farm and homestays and a variety of historical sites to be visited, mostly located around Mount Somers.

Several fishing guides run trips to the upper Ashburton lakes, a small company operates white-water rafting through the Rangitata Gorge, and another offers ecotourism experiences (see page 137).

Mesopotamia Station provides accommodation and can arrange hunting, fishing, horse-riding and rafting expeditions. On the north bank of the Rangitata, Mount Potts has accommodation for skiers.

The remainder of this chapter examines in more detail some of the tourism activities on offer in the central Canterbury high country.

Ryton Station

Ryton covers nearly 15,000 hectares of spectacular mountain country about an hour and a half's drive from Christchurch. While continuing to run a large flock of sheep, the owners have developed a sophisticated home-stay facility, particularly suited to families. Accommodation includes seven ensuite chalets, each sleeping two people, as well as a self-catering lodge that sleeps 16, and three holiday houses. All meals can be provided.

Among the range of activities offered by Ryton is the opportunity to observe farming routines such working dogs and drafting, shearing and dipping sheep.

Ryton Station farmer Karen Mears and assistant.

Recreational opportunities include four-wheel-drive tours in the mountains (including a 'station to station' journey to Central Otago), horse riding, mountain biking, jetboating and windsurfing on Lake Coleridge and, in winter, ice skating and skiing.

Ryton Station has developed the tourism aspect of its business to such an extent that it attracts over a thousand visitors each season, many of them from overseas and often as tour groups. As an economic venture, tourism contributes considerably more to the station than sheep farming.

Ryton Station homestead with Lake Coleridge in the distance.

Terrace Downs High Country Golf Resort

The construction of this resort in the late 1990s marked a new phase in high-country tourism. It provides luxury accommodation alongside a golf course created high on the north bank of the Rakaia River. The first sight of immaculate greens, golf carts, sand traps, water hazards and large stone buildings, all nestling into a dramatic landscape, comes as something of a shock to those who recall high-country recreation as revolving around a few dilapidated tramping or deerhunters' huts.

Terrace Downs Resort was built on what was once part of Bayfields Station and occupies a series of large terraces that drop steeply down to the Rakaia Gorge. Its greywacke stone buildings blend into landscaped grounds and, as the resort's lyrical publicity says, are set 'against a backdrop of snow-dusted Alps and golden-tussocked foothills that are a hunting, skiing and fishing playground'.

The resort's golf course is regarded as one of the best in New Zealand, though it is not for the fainthearted when a strong nor'wester roars down the

View from Terrace Downs Resort over the golf course to Mt Hutt.

valley. The main building contains bars, a restaurant, conference and function facilities, spa pools, shops and beauty-treatment facilities as well as the golf club-house.

Thrills on the rivers

Few activities better evoke the modern era of tourism in New Zealand than jetboating through the gorges and along the braided channels of a high-country river. Bill Hamilton, the inventor of the jetboat, trialled his first craft on the Waimakariri in the early 1950s. It proved ideal for travel in the shallow, shingle rivers of Canterbury, and by the 1960s the New Zealand Jet Boat Association had been formed. Enthusiasts were soon experiencing the thrill of exploring the Waimakariri Gorge and travelling as far upriver as they could go.

Two companies operate trips up this river, with one leaving from the gorge bridge while the much larger Waimak Alpine Jet is based further upstream, near the mouth of the gorge. The latter operates every day, with three 14-seat boats, and when cruise liners are berthed at Lyttelton patronage booms and continuous trips are made. In 2004 this company carried some 20,000 passengers.

Jetboats in the Waimakariri Gorge.
Courtesy Waimak Alpine Jet

The one-hour journey covers about 15 kilometres, travelling between near-vertical cliffs almost all the way. The trip through this otherwise inaccessible canyon provides an opportunity to admire the spectacular Staircase railway viaduct. The outing can also be combined with the half-day High Country Safari (a four-wheel-drive exploration of Torlesse Station) or as a full-day trip, returning from Arthur's Pass to Christchurch aboard the TranzAlpine train.

The name Waimakariri – 'cold rushing water' – aptly describes this massive river's turbulent power.

Conquering the mighty Waimakariri

Jetboats and rafters now mostly handle the Waimakariri Gorge with ease, but the pioneering trips through this section of the river were dangerous ventures. The first known travellers to conquer the gorge were the doughty Marmaduke Dixon and George Mannering, who, in one day in 1889, paddled light canoes from the Bealey down the Waimakariri to where the river emerges onto the plains. Dixon then continued downriver to Christchurch and attended a ball in the evening! This excerpt from Mannering's description indicates the perils of their jaunt:

> We could see nothing before us but a high wall of bare rock, with a mass of foam at the foot, and could not even tell until we were close up which way the river turned, to the right or left. We soon began to gain confidence though, and by experience found the safest way to get through the rapids was to keep to the edge of the current on the inside, and between the current and the whirlpool which almost invariably accompanies every rapid near its foot . . .
>
> Shortly after entering the gorge I had the misfortune to have a small hole knocked in the bottom of my canoe by a sharply pointed stone and with the water coming in we were forced to stop for repairs which we effected with some red lead and a bunch of tussock jammed against the break from inside the timbers . . .

It was not until the 1920s that the gorge was tackled again, by gung-ho voyagers in a variety of makeshift craft. In one of these trips a pair of travellers drowned, including George Carrington, the force behind establishing organised tramping in Canterbury.

Fluctuating flows demand that jetboat crews are thoroughly familiar with every rock in the Waimakariri Gorge, which occasionally cannot be negotiated when the river is in full flood, raising the water level by up to six metres.

The Rakaia Gorge Scenic Jet leaves from below the bridge on State Highway 72 and offers a short but very spectacular journey before the river widens out in the upper Rakaia Valley.

On all four wheels

Several companies run four-wheel-drive high-country tours for passengers or provide hired vehicles for enthusiasts. These are generally based outside the area and vary from half-day trips to journeys of up to a week. The longer safaris usually follow a north–south route between the large rivers, and are hosted each night on one of the high-country properties.

Among the companies that run shorter tours is 4 x 4 New Zealand, based in Geraldine in South Canterbury, which has a fleet of four large vehicles and caters for more than a thousand visitors each year. Their trips include visits to sheep stations and *Lord of the Rings* locations, and specialised photographic and ecotours.

Back Country New Zealand, operating out of the Methven district, aims at the upper end of the market, especially American tourists, and their fleet includes helicopters to carry clients into more remote areas.

A customised four-wheel-drive vehicle crossing the upper Rangitata. The valley in the background was used as the location of Helms Deep in the *Lord of the Rings* films.

Courtesy 4 x 4 New Zealand

Helicopters have opened up previously inaccessible areas to those able to pay for this luxury. They are used on scenic flights, fishing and hunting safaris, and to carry skiers, snowboarders, mountain bikers and bungy-jumpers into the back country.
Courtesy Back Country New Zealand

A fishing safari client displays a large salmon.
Courtesy Back Country New Zealand

Guided fly-fishing and wilderness tours are popular Back Country packages, and the company also has kayaking, mountaineering and heli-skiing options.

Safari hunting

Guided safari hunting attracts those willing to pay very large sums of money to shoot trophy animals either on private game estates or in the mountains, particularly at the head of the Rakaia Valley. There are several properties in the Rakaia Gorge/Methven area that specialise in providing accommodation and guides for well-heeled hunters.

High Peak Game Estate has been extensively developed by James Guild, who in the 1980s set aside 4,500 hectares of the more rugged hill country of his property as a game estate. This has been stocked with red deer as well as some wapiti and fallow deer. Over the years the quality of the famous Rakaia red deer stags has declined in the wild because of excessive trophy hunting, but the High Peak herd was improved initially by importing English stags and, more recently, by a careful breeding programme, so the trophy animals on this game reserve are regarded as among the best in the world.

The Department of Conservation issues concessions to a number of hunting guides to take clients by

Trophy hunters with tahr (***left***) and chamois (***right***) shot on guided helicopter trips to remote ridges in the Rakaia headwaters.
Courtesy Back Country New Zealand

helicopter into known deer, chamois and tahr habitats, particularly in alpine country at the head of the Rakaia Valley. A lodge on the remote Manuka Point Station specialises in hosting such expeditions.

Ecotourists photographing a vegetable sheep on the slopes of Mt Hutt.

Courtesy Tussock & Beech Ecotours

Ecotourism

As interest has developed in understanding and protecting New Zealand's natural heritage, some tourism operators in the Canterbury high-country area have incorporated ecologically based trips and education in their programmes.

Wilderness Lodge Arthur's Pass is run by Gerry McSweeney and Anne Saunders, long-time conservation activists. Their guests are offered audio-visual talks, a variety of guided walks into the surrounding wilderness areas and the opportunity to learn about innovative 'green' farming practices.

Tussock & Beech Ecotours, based at Staveley, a small settlement at the edge of the mountains between Methven and the Ashburton Gorge, provides accommodation in the historic Ross Cottage and one- to three-day trips in the surrounding district. Many of their activities are based on client requirements and range from courses for students studying ecotourism to guided tramping and climbing, bird watching and visits to wildlife refuges. Guidance is also given in nature photography.

Seeking enjoyment strenuously: cycling across scree in the Craigieburn Range. Note the wilding pines, which are seeding and becoming a pest at this high altitude.

Courtesy Dave Mitchell

Challenging the terrain

The central Canterbury high country is an ideal playground for devotees of the increasingly popular sport of mountain biking. Possibilities range from a Sunday afternoon jaunt along ski-club roads in the Craigieburn forest to a strenuous day pedalling and scrambling across seemingly impossible tracks and steep scree slopes to reach an alpine hut.

Mountain biking is generally an informal activity, but some clubs organise competitive events using tracks in the forest and through riverbeds.

Pain for gain – Coast to Coast

While some of the new types of high-country tourism are based around comfortable access and the use of expensive technology or guides, one notable exception involves extreme physical endurance and the ability to push through the pain barrier.

The iconic Speights Coast to Coast multisport competition has been staged in February every year since 1983. This 243-kilometre race from Kumara Beach on the Tasman Sea to Sumner Beach on the Pacific Ocean includes a gruelling section across the

Main Divide and into the upper Waimakariri Valley. Competitors run, cycle and kayak through a series of stages, either as individuals or in teams of two. Most take two days to complete the event, but elite Coast to Coasters tackle the entire course in a single day.

After running and cycling from Kumara to Otira, competitors face a gruelling 33-kilometre boulder-hop up the Deception River, over the 1,100-metre-high Goat Pass and down the Mingha River to Klondyke Corner near the junction of the Bealey with the Waimakariri. The two-day competitors camp here for the night before cycling 15 kilometres then kayaking 67 kilometres down the Waimakariri River, mostly through the grade-two rapids in the gorge. Survivors complete the course with a 70-kilometre cycle ride across the Canterbury Plains to Sumner.

Competitors picking their way through boulders on the alpine leg of the Coast to Coast.
Paul Farrow

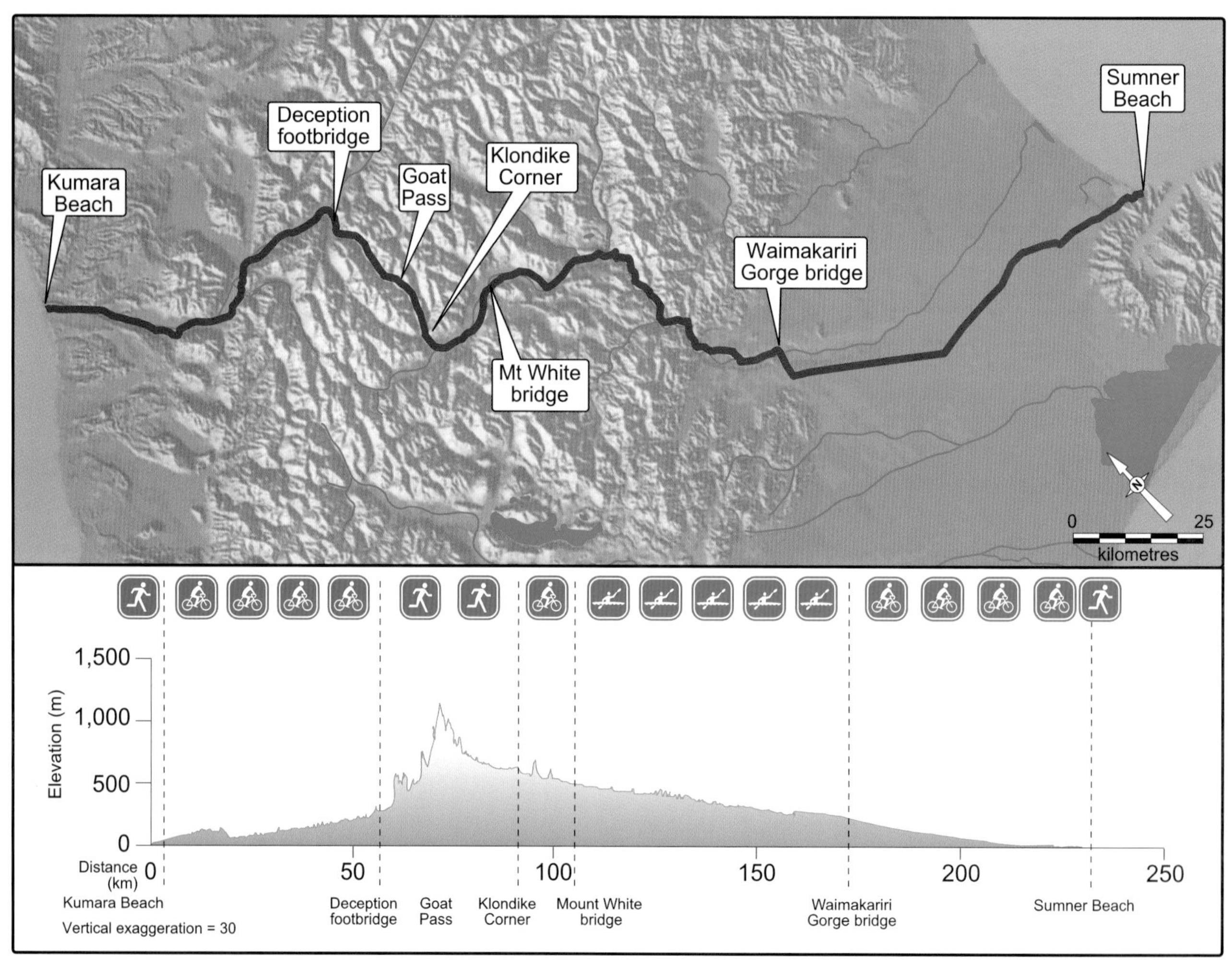

Stages in the Coast to Coast multisport race.
Diagram by Tim Nolan

This event has been organised by Robin Judkins since its inception and attracts at least 800 masochistic competitors, each with their support team. For the first four years it was a rather low-key event with doughty 'mountain men and women' participating. In 1987 the race was discovered by the media and publicised widely, in part because of the dramatic scenery along the route. Suddenly, athletes from other sports were attracted, high-tech cycles and kayaks were developed, serious training and diet regimes were adopted and the Coast to Coast became the benchmark by which other multisport events are judged, both in New Zealand and overseas.

Not surprisingly, the alpine weather frequently plays a part, with almost half the races affected by stormy conditions, particularly in the mountain run and on the kayak section in the Waimakariri Gorge. Safety has become an important aspect of the race planning. Helicopters oversee the Main Divide crossing and a flotilla of jetboats accompanies the kayaks. Despite these precautions, each race is packed with drama, with cycle crashes and capsizes aplenty.

Cycling from Klondyke Corner to the Mount White bridge on the upper Waimakariri Valley stage.
Paul Farrow

Just for the record

Fastest time by an individual in the Coast to Coast: 10 hours 34 minutes 7 seconds – Keith Murray (1994).

Fastest time by woman competitor: 12 hours 9 minutes 26 seconds – Andrea Murray (1997).

Slowest time: 24½ hours.

Average age of competitors: 36.

Oldest competitor: 76 years.

Youngest competitor: 15 (the age limit is now 18).

Most wins: Steve Gurney (9) and Kathy Lynch (5).

Other adventure activities

Alongside commercial ventures, the central Canterbury high country is the venue for a number of lower-profile adventure activities. Rock climbers scale and abseil among the limestone outcrops at Castle Hill, and a nearby recreational reserve features a nearly 600-metre tunnel carved by Cave Stream that can be negotiated in suitable conditions by those prepared to wade in the dark through cold, waist-deep water.

At Arthur's Pass an enterprising company promotes an outdoors wilderness experience featuring a guided walk up an alpine valley to camp overnight in a tent. At the other extreme, tourists can be chauffered in a four-wheel-drive vehicle high onto a mountain and served champagne while watching the sun set.

We have seen that a number of new forms of tourism have been introduced to the high country of central Canterbury. However, apart from the brief, intense media coverage of the Coast to Coast multisport race, the region is not yet showing many obvious signs of these developments.

An occasional helicopter may clatter overhead, and jetboats sometimes roar through river gorges, but generally the valleys are vast enough to absorb these new activities with a minimum of fuss. But for how long will this be the case? Will more luxury resorts like Terrace Downs spring up throughout the high country? Will the two or three little villages of holiday homes grow and sprawl over the hills as they have done in parts of Central Otago?

Continued expansion of high-country tourism potential is inevitable, but it is crucial that this be well planned so that the area's main assets – its scenic grandeur, natural values and a liberating sense of removal from the hustle and bustle – are retained.

CHAPTER TEN

Conservation and land tenure

Looking after the land

A look at the map of the original pastoral leases (see page 80) indicates how quickly the high country of central Canterbury was taken up as pastoral runs by squatters. Compare that with the map opposite, which shows the way the region was being managed in 2006, and it is apparent that changes in land use over 135 years have been immense. Most noticeable is the large area – close to 40 per cent – coloured in green, which is all Crown land that is now unallocated for pastoral leases and is mostly administered by the Department of Conservation (DOC).

The gradual establishment of conservation land is better understood if first we examine the ways that the original pastoral leases were developed and have changed, along with public attitudes towards the high country.

Early exploitation

When the Canterbury Association began leasing blocks of the high country to the first squatters in the 1850s there was no awareness of the conservation or sustainable land-use issues that concern us today. Why should there have been? Here was such a vast, untapped resource that it must have seemed inexhaustible. The leasehold conditions were very short term, which did not encourage enduring care of the land, and most pastoralists came to the colony with the prime objective of making quick money.

Small areas of lower and better farming country in the valleys and foothills were soon freeholded, but in most of the hill and mountain country a system of short-term leases from the government, with a right of renewal, evolved and remained in place for nearly 90 years, until 1948.

This system applied to all of the high country of the South Island, and while central Canterbury represents less than 20 per cent of such land, it can be regarded as typical and most of the material in this chapter applies equally to all back-country pastoral runs. The only major exception here to the government leasing arrangement is that of an area of land owned by the University of Canterbury (see page 142).

For many years the original runholders and their successors endeavoured to graze as many sheep as possible on the native grasses of the tussock lands. They had no previous experience of managing this type of country and learned to cope by trial and error. As described in earlier chapters, many areas were heavily grazed, regularly burnt and later invaded by exotic plants and animals, some of which became pests. There was no incentive to improve the land; in fact, on many properties the only fencing was one separating the 'back' (summer grazing) country from the 'front' country. This severely limited grazing-management options.

Degradation up to 1950

As early as the 1880s signs of degradation of the land were starting to appear and stock numbers began to decline. Burning and heavy grazing resulted in tall tussocks being replaced by short tussocks, invasion by weeds and the rapid spread of rabbits in the lower and drier areas. Soil erosion became evident on the higher country, and forests were opened up by deer destroying the undergrowth.

By the 1920s and '30s some runholders began to examine and modify their management practices. A booklet published in 1992 by the South Island High Country Committee of Federated Farmers included a succinct summary of past practices: 'What we had up to 1948: Short-term lease. Few management constraints, if any. No permanent right of renewal. Land abuse. Exploitation.'

New Zealanders in general were also changing their perceptions of the natural landscape and the effects humans were having on it. In 1923 the Royal New

Zealand Forest and Bird Society was established. At first its interest was narrowly focused on protecting rare birds, but this was the beginning of an increasing awareness of natural values and the threats they faced. A few years later, Arthur's Pass National Park was established (see page 147)

A period of stability: 1948–72

The 1948 Land Act set in place a system whereby Crown-owned high-country land was leased to farmers on a 33-year term, with perpetual right of renewal. The lease gave rights to grazing and the control of public access on that holding. In return, the farmer had to obtain Crown consent in order to modify the land – in particular, to burn vegetation, to cultivate or bulldoze the soil or to plant trees. The programme was administered by the Lands and Survey Department, which employed several pastoral land officers to liase with runholders on behalf of the Crown.

This system generally worked well for over 20 years. The early land officers were well accepted by the runholders and had similar objectives to theirs:

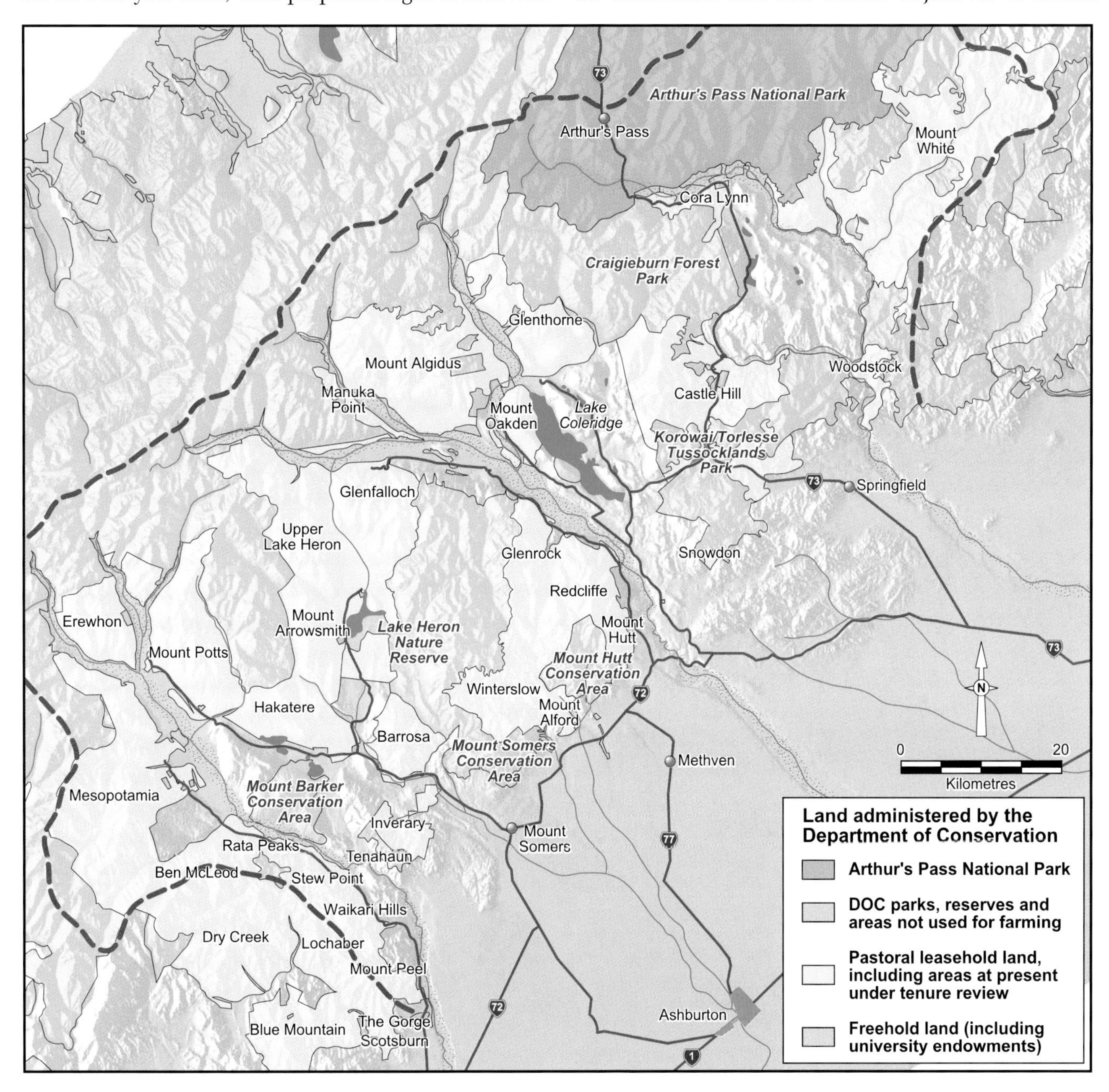

Central Canterbury land management in 2006. Because of the ongoing tenure-review process, this map is a snapshot of the situation at the time of writing.

Cartography by Tim Nolan

the first chief PLO set up a system of low rents in exchange for light stocking, and the runholders, with a real sense of long-term ownership, could look to the future with confidence.

Eventually the runholders established the South Island High Country Committee of Federated Farmers and began speaking with a united voice. As concern for the welfare of the land increased, various other groups were set up. Rabbit boards made concerted attempts to reduce numbers of that pest, and catchment boards collected information and advised and assisted runholders in the management of river flows and, particularly, the erosion of the higher and steeper slopes.

Although some scientific research on the vegetation of the high-country had been undertaken for many years (most notably by Dr Leonard Cockayne), this intensified in the 1940s. Much of the study was based at what would become Lincoln University, where the Department of Agriculture established a grasslands research section, and in 1961 the Tussock Grasslands and Mountain Lands Institute was formed. Notable leadership was provided by Lance McCaskill and then by Professor Kevin O'Connor.

The 25 years from 1948 were a relatively stable period in the high country. Runholders were able to obtain loans more readily and thus invested more in the land. Aerial topdressing was widely used to raise the quality of pasture, fencing and other improvements were carried out, and conservation values tended to become more important, particularly in the controlling of soil erosion and pests. Relations with the Crown were generally comfortable and a solid body of scientific research on the management of the tussock grasslands was built up. The general public increasingly came to appreciate the natural values of the high country as they used the area for tramping, hunting and fishing.

In the 1950s some of the smaller properties in the easier country to the east, particularly in the Rakaia Valley, were subdivided into smaller farms for the settlement of returned servicemen. For some of these, and other smaller properties, the 1970s proved to be a particularly good time as many farmers benefited from government subsidies.

By the mid-1970s New Zealanders' perceptions were once again beginning to change about the high country and its use.

Developing conflicts of interest: 1970s–90s

During the latter part of the twentieth century runholders throughout the South Island felt increasingly under pressure, largely as a result of growing interest in the high country by other groups, who have exerted considerable influence on the government. A three-way tug-of-war developed among runholders, conservationists and recreational users, and led to changes in government policies.

University of Canterbury endowment land

While the bulk of our region's high-country runs have until recently been made up of land leased from the government (known by the original Canterbury Provincial Council as 'Waste Lands of the Crown'), there are also significant parts of half a dozen properties that are leased from the University of Canterbury.

In 1873 William Rolleston, Superintendent of Canterbury (and a very early runholder at Mount Algidus), proclaimed that some 116,000 hectares of the province's high country was to be reserved for the then Canterbury College School of Agriculture and School of Technical Science 'as endowments for the promotion of superior education'.

Nearly 90,000 hectares of this endowered land were in the catchments of the area discussed in this book: around Cass in the Waimakariri Valley, near Lake Coleridge in the Rakaia catchment and in parts of the middle Rangitata Valley.

Over the years some of this property has been sold to private runholders (mainly in the Rangitata), but there are still about 40,000 hectares of university endowment land in the region.

Of the approximately 25,000 hectares in the Cass area, more than half is on the Flock Hill property and about a quarter on Craigieburn. Grasmere has a lease on a small amount, and most of the remainder is set aside as a scientific reserve around a university field station on Sugarloaf Hill at Cass.

Further south there are about 14,000 hectares of university land around Lake Coleridge. Most of this is on Ryton Station, with smaller amounts on Lake Coleridge and Acheron Bank.

This land continues to be administered under similar terms to Crown leases, and at the time of writing there are no plans to alter this situation.

Obtaining information about erosion

When, in the 1950s, it became apparent that there was a woeful lack of hard evidence about the effects of burning and grazing of mountain vegetation, the North Canterbury Catchment Board set about addressing the problem. This 1953 photo shows catchment board officers recording the vegetation cover in a transect running down a slope in the Castle Hill Basin.

Statistics from many transects were collected over a number of years and the vegetation changes were noted in a series of fenced-off sample plots at high altitude, particularly near the summit of Porters Pass.

Conservationists' viewpoint

'The high country of the South Island has a special place in the psyche of New Zealand . . . it blends into an iconic landscape of shingle riverbeds, tussock-floored basins, lakes, wetlands, and the ranges themselves.'

Few would disagree with this view presented by the Forest and Bird Society, and today most New Zealanders are familiar with high-country landscapes and are proud of the way they are used in publicity material to attract overseas tourists, and as backdrops in a number of television and print advertisements.

Just as important as this iconic view is the fact that, to scientists and conservationists, these landscapes – particularly the tussock grasslands – form rich ecosystems that are the habitat of a surprisingly large range of species of unique plants and animals. Unlike other distinctive New Zealand ecosystems, however, very little of the tussock grasslands and their associated communities had been set aside as conservation land to be protected until as recently as 2001.

Some conservationists claim that, under continued grazing, the tussock grasslands are being changed and degraded, so that at higher altitudes soil erosion increases, the entry of exotic weeds and pests is made easier, and therefore many indigenous species are unable to compete and are becoming endangered. Thus, they argue, grazing stock – on the higher country at least – should not be allowed.

Recreational groups' viewpoint

By the 1970s the Federated Mountain Clubs plus hunting and fishing interests were expressing concern about restricted access rights to Crown leasehold land. In the past, New Zealanders regarded wilderness areas as part of their heritage, to be roamed over as they wished. Relations between most runholders and recreational users had generally been good, and access was usually granted to those who asked for it. Fears that this tradition was under threat arose when several high-country properties were sold to overseas interests who, it was felt, may deny right of way.

Runholders' concerns

Changes in public attitudes towards the high country eventually persuaded the government to impose more management constraints on Crown land runholders. Several agencies became involved, each with their own, at times conflicting, objectives.

How have these pressures and changing attitudes affected the runholders? In a 2004 article in the *New Zealand Listener*, 'typical' high-country farmers were described as an 'independent, wilful, resourceful, staunch, exasperating little group of families who, generically at least, have always been there'. If this were an accurate description, it would not be surprising if they were unhappy with increasing government 'interference' and the shift in public perception, but the reality is that while some runholders just want to get on with their traditional job of raising sheep, the majority are well in tune with the changing times, have already adopted new farming methods and are looking to diversify and, often, to embrace tourism.

A good deal of anger has been expressed by some farmers over accusations of exploitation, a criticism that may have had validity in the distant past, but the present runholders believe that they understand their land and maintain it well. Increasing frustration over government constraints on what farmers can do on their leasehold land has been accentuated by the desire of many to develop the land in ways other than for pastoral grazing.

The tenure review process

By the late 1990s it seemed that the views on the high country held by the three competing interests – economic farming, recreational access and conservation – had become irreconcilable. The government responded by passing the Crown Pastoral Lands Act 1998, which introduced tenure review and involved the following process:

A gully on the road to the Mt Olympus skifield, near Lake Coleridge. This sort of country is unsuitable for the grazing of sheep and could be taken over as reserves by the Department of Conservation. Compare this with the photo on page 106 of High Peak, which is a fully freehold property that has been extensively developed. Although neither area has undergone tenure review, they indicate the possibilities intended by this process.

- Any lessee may ask Land Information New Zealand (LINZ) to review the tenure of their lease. (LINZ had replaced the Lands and Survey Department as controller of pastoral leases.)
- LINZ consults with the Department of Conservation (DOC). A detailed inspection of the land is made, other interested parties such as Maori, and Fish & Game are consulted and public submissions invited.
- A proposal is designed to divide the property into freehold and conservation areas, with public access routes over the freehold land. If this is acceptable to the lessee, negotiations are carried out to complete the process.

Tenure review was intended to be a win-win situation for all parties. Farmers could freehold the most productive part of their property, which could then be managed with no restrictions, planted in forest or sold. The advantage to the public good is that the more fragile, uneconomic land will be protected from damage by stock and come under DOC administration.

Tenure review in operation

After six years about two-thirds of the 300 pastoral leases in the South Island have passed through or are undergoing tenure review. (The ratio is similar in the central Canterbury high country.) In many cases the reviews have been welcomed by runholders and conservationists alike, but in some instances they have been contentious.

For many runholders, the tenure review option has been seen as an opportunity. Although the process can be slow to implement, it has been relatively successful for runs that are a clear mix of mountains and river flats, and thus quite easily subdivided. In such cases the hoped-for win-win situation seems to be working and freeholded land is being diversified.

But the concept is a concern for some farmers, particularly those who rely heavily on extensive areas of mid-altitude tussock hill country for grazing. Conservation interests tend to consider that such marginal land should be included in the area to be protected. The affected runholders believe that their operations would not be viable without large areas of this country, and that retaining the old grazing leases, with their low rental and security of tenure, may be a better option for them.

There are claims that DOC will not be able to adequately manage the huge increase in conservation land, and that the high country will be overrun with

View of north-facing slopes of the Palmer Range on Double Hill Station. Following tenure review, most land above the treeline – about 1,000 metres – is passing into the control of the Department of Conservation, while the lower slopes and flats are being freeholded. The effects of topdressing on this land are obvious.

weeds and pests. Indeed, many runholders feel that they are already successfully managing and conserving the land, and some consider that multiple use with light grazing may often be a better solution than the complete closure of land with high conservation values, particularly where weed control could become a problem.

Although concerned farmers are told that there is no compulsion to undertake tenure review, a few are sceptical and suspect that when the majority of leaseholders have completed it, those remaining will be put under pressure to do so.

On the other side of the coin, the High Country Coalition has brought together Forest and Bird, Federated Mountain Clubs, Public Access New Zealand, Federation of Fresh Water Anglers and the Council of Outdoor Recreation Organisations into a powerful lobby group to protect conservation and recreation interests. While very pleased with the outcome of some tenure reviews, the coalition is dismayed by others. Conservation groups in particular believe that farmers are getting a somewhat greater share of the land than is being allocated as reserves, and that they are often getting freehold title at a very low price. An extreme view is that the high country is being 'privatised for peanuts' and that the government is giving away much of our high-country heritage. This opinion is strongly disputed by the High Country Accord, which was set up in 2003 to represent the interests of most of those farming Crown pastoral land in the South Island.

Access is the main problem for recreational users. The land freeholded to the farmers becomes private property and it is felt that access is less likely to be granted to trampers, hunters and fishers.

The role of government agencies

What the government wishes to do, according to a DOC press release of 1 June 2004, is to 'add important landscapes, ecosystems and public access/recreation opportunities unique to the high country to the conservation estate via a network of high country parks and reserves'.

While LINZ runs the tenure review process, DOC is a major adviser and has the role of managing the resulting conservation land.

DOC's statutory responsibility is to protect native species, foster recreation and manage cultural and historic heritage on the conservation estate. As the amount of land coming under the department's care increases, weed control has become a particularly vexing problem, especially the eradication of wilding pines, although the government has earmarked more funding for this purpose. One of the major benefits of tenure review, in DOC's opinion, is that public access to high-country parks and reserves will be improved under its administration.

As the tenure review process has continued, New Zealanders in general are becoming more aware of what is happening in the South Island high country. It is becoming clear that the process is slow and complex, and its implementation continues to have both supporters and vehement critics.

A common concern was expressed in a *New Zealand Herald* article in mid-2006 that quoted Alan Jolliffe, chair of the Canterbury-Aoraki Conservation Board, as saying: 'Tenure review is disposing of the crown jewels of the South Island by stealth.'

Apprehension is also being expressed that iconic natural landscapes will be disfigured by tourist developments on newly freeholded land.

Double Your Luck in the High Country, Don McAra, 2006.
Of this painting the artist writes: 'There is a topical concern over the future of high-country runs, such as this one at Flock Hill Station, which has diversified successfully. Others, however, once freeholded may be at risk of being put out of the reach of New Zealanders.

'The wheel of fortune for our high country inheritance may change in ways that we may come to regret.

'However, this painting also commemorates the artist waking up at Flock Hill to enjoy the light of a new day stealing over this historic station, with the memory of many happy days tramping in the area.'

Courtesy Don McAra

These fears appear to be less of an issue in the central Canterbury high country than in Central Otago and the Mackenzie Basin. Nevertheless, evidence of the results of these changes is beginning to appear. The valley floors are becoming much more intensively cultivated and green pastures are replacing the hardy sweet vernal and browntop among the tussocks on the lower hills. There is also renewed building in villages such as Castle Hill and Bealey, and several properties are subdividing sections for holiday homes.

Overseas ownership

The issue of overseas ownership of areas of the South Island high country has been thrust into the spotlight as a result of the increased amount of freehold land created by the outcome of tenure reviews. This region tends to be of particular interest to overseas buyers because of its superb environment and isolation, combined with the comparatively cheap price, by their standards, of any land coming onto the market.

While worries that overseas owners may prohibit public right of way to heritage land have been borne out in one or two cases, in other instances not only has access been maintained but the new runholders have invested significantly in preserving the natural environment.

In the central Canterbury high country the issue of overseas ownership at present affects only three or four properties.

Tenure review in central Canterbury

Each case of tenure review is different, with many complex issues to be worked through, so the LINZ goal of completing the process by 2008 is perhaps optimistic. As of 2006, only about half the properties in our region had completed review and the proportion of land going to the conservation estate has varied greatly. It is likely that the full negotiation programme will result in a network of conservation parks and reserves (including Arthur's Pass National Park) covering over 50 per cent of the central Canterbury high country.

The greatest changes to land tenure have taken place in the Waimakariri catchment, where the nine runs that were originally taken up are now almost unrecognisable. The present land use is a microcosm of almost all of the various possibilities that have evolved over the past 150 years and close to two-thirds of this area now consists of conservation park.

The three original stations encountered on State Highway 73 after leaving Springfield – Ben More, Brooksdale and Castle Hill – have all been subdivided. Most of the higher country is now incorporated into the Korowai/Torlesse Tussocklands Park and the remaining freehold land on the flats is being developed for much more intensive farming.

Further up the highway, Flock Hill and Craigieburn Stations are largely University of Canterbury leasehold land, and both have incorporated tourist activities into their operations. Cora Lynn is undertaking tenure review and, like Grasmere, has developed upmarket tourist accommodation and activities. Only the remote Mount White remains mostly leasehold land farmed as a traditional pastoral run.

The properties in the Rakaia Valley have also changed dramatically, in two very different ways. Almost all of the land between the river and Lake Coleridge, the terraces further south and the High Peak Basin have been subdivided into freehold properties that have been improved for farming.

The more mountainous country further up the Rakaia has a variety of tenures: Ryton is partly leasehold and partly Canterbury University endowment lease; Glenthorne and Manuka Point are currently in the process of tenure review; and Peak Hill has completed negotiations, with about a third of the property becoming conservation land.

On the south bank of the Rakaia, Double Hill has completed review and some 60 per cent of the original property, including the higher, north-facing slopes of the Palmer Range, has joined the conservation estate, leaving the remaining lower slopes and river flats as freehold property. Tenure review has been concluded at Glenariffe and is under way on Redcliffe, Glenrock and Glenfalloch Stations.

Further south, change has been rather slower, partly because of the more difficult access to the runs in the Lake Heron Basin and the upper Rangitata. A good deal of higher mountain country has been retired from grazing and passed over to DOC, and part of Clent Hills has recently been made into a reserve. Tenahaun has completed the process and properties under review include Mount Alford, Mount Potts, Mount Arrowsmith, Barrosa and Mesopotamia.

The establishment of conservation parks

During the last 75 years New Zealanders' interest in the natural world and in conservation has increased to the extent that it has led to the establishment of a series of areas to be controlled by the government for the protection of the indigenous species. The first area set aside in the Canterbury high country was Arthur's Pass National Park, followed more recently by several forest and tussockland parks and reserves. By 2000 there was a total of almost 200,000 hectares of conservation parkland set aside in our region.

The total area of land administered by DOC is much more than this because the department is also responsible for the large strip of alpine country on the eastern slopes of the Main Divide, stretching about 75 kilometres from Arthur's Pass National Park to Mt D'Archiac at the northern end of Aoraki/Mount Cook National Park. This land has never been allocated for pastoral leases.

In recent years some large portions of Crown land have been retired from grazing, including the higher parts of the Arrowsmith Range, west of Lake Heron, and Mt Harper, on what was once part of Mount Possession Station, as well as a number of smaller areas.

When all of these areas are included, the conservation estate covers over 40 per cent of all of the high country in central Canterbury, and this proportion will increase further as the tenure review process continues.

Arthur's Pass National Park

New Zealand's third national park (after Tongariro, 1887, and Egmont, 1900) was established in 1929, but as early as 1901 the notable botanist Leonard Cockayne was instrumental in persuading the government to gazette 43,800 hectares of the Bealey and Otira area as suitable for a park. He spent a great deal of time exploring in the region and considered it 'a fine example of transalpine flora in transition', very worthy of preservation.

The lower slopes of Arthur's Pass National Park are covered with mountain beech forest and, at higher altitudes, alpine ecosystems are easily accessible from the road at the top of the pass.

Unfortunately, by the time the park was established some degradation had occurred. In 1890 much bush was burnt in a fire started by a railway survey party, and in the 1920s another fire and tree cutting by railway tunnellers caused further damage. The more accessible vegetation was soon under threat as the flood of visitors on excursion trains returned home with armfuls of local flora.

After the national park was established, and a board appointed to manage it, there was little money available during the Depression to devote to improvements, and efforts were largely confined to the area around the village. Elsewhere, almost all of the development of tracks and hut construction was done by tramping clubs, a situation that continued for many years.

Additions were made to the park in 1931, 1938 and 1950, increasing the total area to almost 100,000 hectares. In 1950 several innovative people were appointed to the park board, more money was made available and two full-time rangers were soon appointed. The administration of the park is now under the control of DOC, operating from an attractive and informative headquarters in the village.

Craigieburn Forest Park

The second area to be recognised as being worthy of conservation status was the mountain beech and alpine area of the eastern Craigieburn Range, which was designated a forest park in 1967. Additions in 1979 and 1984 took in parts of the western side of the range and extended into the catchments of tributaries of the Rakaia, including the valleys of the Avoca and Harper Rivers, the surrounding mountains and forests, and also most of the forests on each side of the Wilberforce River. Craigieburn Forest Park now covers 44,000 hectares.

Because access is difficult, relatively few trampers and hunters visit the valleys on the Rakaia side of the park, but the Craigieburn side has easy access from SH73 and the side roads leading to skifields, as well as numerous tracks, ranging from 20-minute to full- day walks. Located within the lower forest is an environmental educational centre, and some 7,000 people visit the area each year to participate in a range of activities, including climbing, walking, mountain biking, skiing, school visits and picnics.

Conservation areas further south

Among several conservation areas administered by DOC in the front ranges between the Rakaia and Rangitata are those on Mt Hutt and Mt Somers. The former includes several reserves covering over 4,000 hectares and stretches from the base of Mt Hutt, through predominantly beech forest in the valley of Pudding Hill Stream, up to the alpine basin where the skifield is situated.

The Mount Somers Conservation Area encompasses the entire 1,687-metre-high mountain. Alford Forest covers much of the lower slopes, with alpine scrub present near the summit. Trampers have a choice of several short walks as well as high-altitude routes, with two huts providing accommodation for overnight treks. The remains of the Blackburn coal mine can be seen on the western side of Woolshed Creek.

The Mount Barker Conservation Area has recently been created as a result of the retirement from grazing of part of Mount Possession Station.

Korowai/Torlesse Tussocklands Park

In 2001 a new phase in the development of conservation land in the high country was launched with the establishment of this park, largely consisting of former pastoral-lease land with the purpose of preserving examples of mountain beech forest and alpine ecosystems to the east of the high country. It includes snow tussock grasslands, shrublands and scree, and is home to a range of distinctive insects and other animal life.

The park was originally proposed by the Forest and Bird Society and was created by combining a series of blocks of land, most from the Brooksdale and Ben More Stations and the old Avoca run. Some of this was Crown land retired from grazing, and some was purchased through the Nature Heritage Fund. The result is a 22,000-hectare park on some of the mountains familiar to those who travel to and from the West Coast on SH73. It includes the main Torlesse Range, most of the Big Ben Range and Mt Lyndon to the west, as well as much of Thirteen Mile Bush.

The highway over Porters Pass and the Lake Lyndon to Coleridge road both pass through the park, so its lower sections are very accessible. There are as yet few tracks, but the region offers considerable recreational opportunities for trampers, cross-country skiers, naturalists and photographers.

An addition to the park will take in much of the higher country of Castle Hill Station following completion of tenure review. This will include the north face of the Torlesse Range and connect Korowai/Torlesse to the Craigieburn Forest Park. Within the Castle Hill area are several reserves established some time ago to protect the distinctive limestone outcrops and the unusual plants that grow among them.

Lake Heron Nature Reserve and Clent Hills

In 2004 the Nature Heritage Fund negotiated with local farmers to purchase about 10,000 hectares of Clent Hills Station in order to create a conservation buffer zone around the wetlands of the Lake Heron Nature Reserve. This area consists of tussocklands with several small lakes, and extends east from the lake to the summit of the Mt Somers Range, with peaks above 2,000 metres. It will not only protect the

A view from the top of Porters Pass, down Starvation Gully to Mt Lyndon. This peak and the ridges on each side of the road are part of the area that in 2001 became the Korowai/Torlesse Tussocklands Park.

wetlands but provide recreational opportunities for tramping, fishing, ecotourism and mountain biking.

The process of preserving natural heritage in the central Canterbury high country will certainly continue. There are already suggestions that the Arrowsmith Range and surrounding country would be a suitable area for the creation of another conservation park. When completed, the tenure review process will result in our region's becoming a mosaic of conservation parks and reserves, mainly in the higher areas, and intensively farmed freehold properties in the valleys and basins. The hope is that diversification on the improved land and the likely increase in tourism activities and holiday-home building will not despoil the magnificent high-country landscape.

The changing face of the high country. ***Left:*** Larches spreading in the Castle Hill village area. A new road is under construction to the left of the village to accommodate more holiday homes. ***Right:*** One of the smaller freehold properties in the Rakaia Valley offering subdivided sections as the 'ultimate lifestyle investment opportunity'.

CHAPTER ELEVEN

The Inspirational Land

High-country art and writing

In this scarred country, this cold threshold land,
The mountains crouch like tigers.

– 'The Mountains', James K. Baxter, 1942

Few people who have experienced the Canterbury high country have failed to be affected by the brooding presence of the mountains as the weather closes in, the vastness and grandeur of the open basins, or the strong colours of golden tussocks contrasting with dark green forests.

Since the time the area was explored this remarkable landscape has inspired artists and writers to record their impressions in paint, poetry and prose. In more recent times the open vistas of Central Otago may have attracted more attention, but the impressive mountains and valleys of the central Canterbury high country have always fascinated creative people. This chapter highlights a few examples of art, poetry and prose inspired by the region.

Early impressions

Many of the first settlers who came to the high country were somewhat intimidated by what seemed an alien environment. Some were moved to record the shifting moods and dramatic weather conditions of the mountains and valleys, together with the challenges of trying to tame the new land. Many of the colonists

Clent Hills, Robert Booth, early 1860s.
Booth was a young man who worked for a year or so for Samuel Butler at Mesopotamia. This painting is of the newly built homestead on the neighbouring Clent Hills Station, a little enclave of tidy domesticity surrounded by tussock-covered hills and with the massive Arrowsmith Range looming in the background.
Courtesy of Hocken Library, Dunedin

Mesopotamia, Captain E. F. Temple, 1884.
Courtesy B. Temple

were well-educated and skilful writers; others were competent at sketching and watercolour painting. They were keen to pass on their impressions and novel experiences to those they had left back home in England. As a result, we have a number of keen insights into the lives of these pioneers as they tried to establish themselves.

Like a number of early settlers, Samuel Butler stayed for only a few years before returning to England, but he found time to write detailed descriptions of the tasks involved in setting up his property. His letters home were collected by his family and published as *A First Year in Canterbury Settlement* (1863). Butler also sketched and painted his surroundings, played a piano that he had transported to his remote homestead, and gathered material for *Erewhon, or Over the Range*, a novel satirising Victorian society but recognisably set in the Canterbury high country.

From 1861 to 1888 John Enys kept a detailed diary of events at Castle Hill Station, and his brother Charles painted many watercolours depicting aspects of the upper Waimakariri area (see, for instance, *The Woolshed at Lake Letitia*, on page 82). Another early artist, William Packe, in 1868 produced the rather romanticised view of the Mesopotamia homestead reproduced on page 78. A few years later, in 1879, Captain E. F. Temple arrived in Canterbury from India. He was a competent artist and his picture of Mesopotamia and the mountains across the Rangitata Valley is more accurate than Packe's. Temple became part of the art scene in Christchurch and painted a number of other works, including one of the Mount Hutt homestead.

Of the many people who wrote detailed accounts of living on the early high-country runs, Lady Barker is the best known for her *Station Life in New Zealand* and *Station Amusements in New Zealand*. L. J. Kennaway's *Crusts: A settler's fare due south* records his short stay in Canterbury (see page 77), and Robert Booth's *Five Years in New Zealand (1859 to 1864)* includes several sketches that illustrate the hardships of pastoral life (see page 79).

A recurring theme of these observers was the settlers' battle with the elements, especially the extraordinary nor'west wind. In 1852 Mark Stoddart, an Australian squatter who took up land at the mouth of the Rakaia Gorge, wrote this vivid description:

> This beautiful spot, however, has peculiar drawbacks of its own – the nor'-west winds, the curse of New Zealand, pour thro' this embrasure of the mountains with a force which must be witnessed to be believed and converts the avenue-like bed of the river into the most howling scene of desolation

> – horsemen are blown out of the saddle, sheep drift before it miles upon miles, cultivation is uprooted and the soil carried bodily away.

Stoddart was so 'blown away' by the nor'wester that he composed a 14-stanza poem evoking its malign presence. Part of 'The Shagroon's Lament' reads:

Among the dreary mountains, far up above the gorge,
There lives a potent demon, ever working at his forge;
A worker at the winds is he, a flatulent old buffer,
And he sends his manufactures down that man and beast may suffer.

I've witnessed all the winds that blow, from Land's End to Barbadoes –
Typhoons, pamperos, hurricanes eke terrible tornadoes.
All these but gentle zephyrs are, which pleasantly go by ye,
To the howling, bellowing, horrid gusts which sweep down the Rakaia.

The little cloud now sailing down is foreman at the bellows.
At Mt. Hutt's base he'll take his place to overlook his fellows,
There's Gust and Puff, and Shriek and Howl, and demons without number;
And they're coming now, with dusky brow, to waken summer's slumber.

The Prince of the Air is roused from his lair,
And howls in his bullying might:
The gravel and dust are now mixed with the gust,
And the demons shriek out with delight.

The best of good fellows can't stand the strong bellows,
That are ever at work on this shore:
So stick where you are, it is better by far,
Than come here and be heard of no more.

By the end of the nineteenth century Canterbury's artists were less concerned with recording the landscape and the human impact on it, and more intent upon appreciating its scenic and 'painterly' values. The few professional artists who had ventured into the high country tended to regard it as a harsh and often barren setting. Petrus van der Velden, for instance, emphasised the dark and forbiddingly 'sublime' in his well-known series of Otira paintings. Only slowly were others such as Margaret Stoddard beginning to see the landscape of the interior in a more agreeable light.

High-country art flourishes: 1930s–60s

By the 1920s the Christchurch art scene had been boosted by the establishment of a School of Art at Canterbury College and a number of emerging artists who began to turn to the high country for sources of inspiration. From the 1930s until the 1960s a steady stream of art was produced that drew upon the region in various ways.

Perhaps the most prolific of the artists active at this time was Sydney Lough Thompson. Although he spent much of his later life in France, Thompson regularly returned to New Zealand to paint. One of his favourite areas of inspiration was the Waimakariri Valley, particularly around Grasmere Station, where he sometimes stayed. He left many fine paintings of this area: *Lake Sarah* is reproduced on page 21, and *Staircase, Waimakariri Gorge* appears opposite.

Thompson has been described as using 'a decorative form of impressionism'. Certainly he viewed his beloved high country in a much more benign way than had artists such as van der Velden. Painting outdoors and using thick layers of oils, he portrayed the region in its most attractive moods, with grand features but also full of sunshine and warm colours. Thompson's fine colour sense and balanced composition made his work very popular and did much to influence outsiders' perceptions of the high country.

Another frequent visitor to the Waimakariri area was Olivia Spencer-Bower. She mainly painted in watercolours, using a muted palate, and her relatively loose style of application conveyed the charming aspects of the Canterbury high country. *Torlesse* is one of her earlier works and is executed realistically. In the later years of her long life as a painter Spencer-Bower tended to work in a very spare way, capturing only the essential details and leaving large expanses of white, giving a feeling of movement.

Cass, by Rita Angus, is undoubtedly the best known of all paintings depicting the Canterbury high country, and in 2006 was voted the nation's favourite work of art. Angus was trained in Christchurch and used the upper Waimakariri area as the setting for many of her earlier paintings. *Cass* exhibits her strong, highly impressionistic style, carefully planned composition, and sense of rhythm. The intense colours, the repeated folds of the hills, the clouds in the nor'west sky and the solitary person and building combine to convey a hot and dry landscape in which the human impact seems dwarfed.

The features of Angus's rural environment and sense of isolation have been said to be aspects of the

***Above:** Torlesse*, Olivia Spencer-Bower, 1938.
Waimakariri Art Collection Trust, courtesy Olivia Spencer-Bower Foundation

***Left:** Staircase, Waimakariri*, Sydney L. Thompson, 1938.
Private collection, courtesy Annette Thompson

***Below:** Cass*, Rita Angus, 1937.
Christchurch Art Gallery, courtesy W. Angus

Above: A Dry Summer – Waimakariri, William Sutton, 1949.
Christchurch Art Gallery

Below: Nearly 60 years later, the bridge has been replaced, Bruce Creek has become Bruce Stream and the flats are sown in grass, but the mountains and the riverbed remain little changed.

beginning of a search for a distinctive regionalism in New Zealand painting.

William Sutton lectured at the School of Fine Arts at the University of Canterbury for over 30 years. A *Dry Summer – Waimakariri* (above) is one of his earlier paintings and, like Rita Angus's work, it offers much more than a simple landscape. The rich ochre colours, super-real treatment of the riverbed stones and the streaky sky contrast sharply with the bridge, an intrusive human construction.

Sutton stands out as a leader in the development of regionalist painting. His later works included a series constructed of stylised rows of dry hills, a device that was followed by a number of other New Zealand painters, including Colin McCahon.

Writing in the high country

New Zealand's best-known poet, James K. Baxter, regarded mountains as strong, dominating forces. His 'Hill Country' (1942) is wonderfully evocative:

White, sky, mountains mount
High: near, terraced, clear
Groined, shouldered
Black-bouldered. Sallow flats lie
Dry . . .
. . . Wind strums over plain
Same dry, high plain:
Thin grass sounds in winds
As winds pass.

Much of the poetry set in the high country was rather less profound and inclined to romanticise life on sheep runs. This urge produced ballads with titles such as 'The Hungry Shearer', 'The Drover', 'Mustering Above the Fog' and 'The Station Dipping'.

Violet Fraser's 'Autumn Muster' (1943) conjures up an idealised Canterbury scene:

In dim outline the hills are seen,
The sky is pearly grey
The morning star shines bright and clear,
Night's pale stars fade away.

A sudden clamour fills the air,
Dogs bark and men call out,
Hooves beat upon the frosty ground,
Dawn's quiet is put to rout.

With whistles shrill and sharp commands
The men are on their way,
Their dogs at heel, towards the hills
They ride, ere it is day.

In red and golden majesty,
The sun now mounts his throne,
The homestead drowses quiet again,
A dog howls – left alone.

Bruce Stronach's 'The Last Run', inspired by an incident while mustering on Mount White Station in 1938, is full of pathos:

He'd fallen over a cliff
And he'd broken his leg.
Just a mustering dog.
And he looked at me, there on the hill,
Showing no hurt, as if he'd taken no ill,
And his ears, and his tail,
And his dark eyes too,
Said plainly,
'Well, Boss, what do we do?
Any more sheep to head?
Give me a run.'
But he'd never head sheep any more.
His day was done.
He thought it was fun
When I lifted the gun.

'The Canterbury Curse', by Jim Morris, returns to a familiar subject:

It's spring once more in the gorges
It's Nor'west time again,
The wind that drives men crazy
And brings the heavy rain.
It floods the shingle rivers
And breaks the women's hearts,
It pays to stock up with tucker
Before the Nor'west starts.

Steadily blowing for weeks on end
It never seems to abate,
And often in the stinging dust
It needs two men to shut a gate.
It strips iron off the woolsheds
And trees snap off like chalk,
And shepherds get a permanent lean
As round the farm they walk.

Now that you've read this tale of mine
You might think I'm a jester,
But any doubts on my sanity
You can blame on the old Nor'wester.

In the 1950s and 1960s life in the high country was popularised by the publication of books by Mona Anderson, Peter Newton and David McLeod, among others. This trend has continued more recently in a number of photographic essays that make the most of the region's scenic marvels, hard-bitten characters and clever working dogs.

Present-day artists

Austen Deans has been painting watercolours since the late 1930s, and much of his very popular work comprises recognisable vistas of the Canterbury high country.

While realistic landscapes continue to be painted and admired, many artists have begun to use the high country as just part of a picture that makes a statement about their concerns for the region.

Don McAra's painting on page 146 celebrates the charms of Flock Hill, but the text he provides warns that such a scene may be threatened as land is freeholded and removed from the public estate.

Above: The Arrowsmith Range, A. A. Deans.
Ashburton District Council, courtesy A. A. Deans

Left: Austen Deans (at right) painting *en plein air* with his wife, Liz Warren, in 2002.
Courtesy Tussock & Beech Ecotours and Austen Deans

Andrew Craig has recently exhibited a series of paintings that focus on the rocky landforms of the high country, particularly in the Rakaia Gorge area (see page 13). These works explore notions of timelessness, the origins of the land and the ways in which its elements are built up and worn down.

In 2004 Michael Hight, an Auckland artist, painted a number of works under the collective title 'Land of Milk and Honey – Images From a Canterbury Winter'. While their central images are of clusters of beehives, they derive much of their impact from high-country settings. In the painting reproduced below, the bare, open basin, the mountain backdrop, patches of snow and the pale winter sun combine to create a feeling of stark loneliness that surrounds and contrasts with the enclosed communities of bees.

Diana Adams is a young Christchurch artist who paints simplified images using bold, contrasting colours. Her *Boulders – Castle Hill* (page 19) and the work reproduced opposite effectively interpret the land by reducing its complexity to simple forms.

Cristina Popovici immigrated to New Zealand from

Beehives Winter Lake Heron, Michael Hight, 2004.
Courtesy Michael Hight

Canterbury Contours, Diana Adams, 2003.
Courtesy Diana Adams

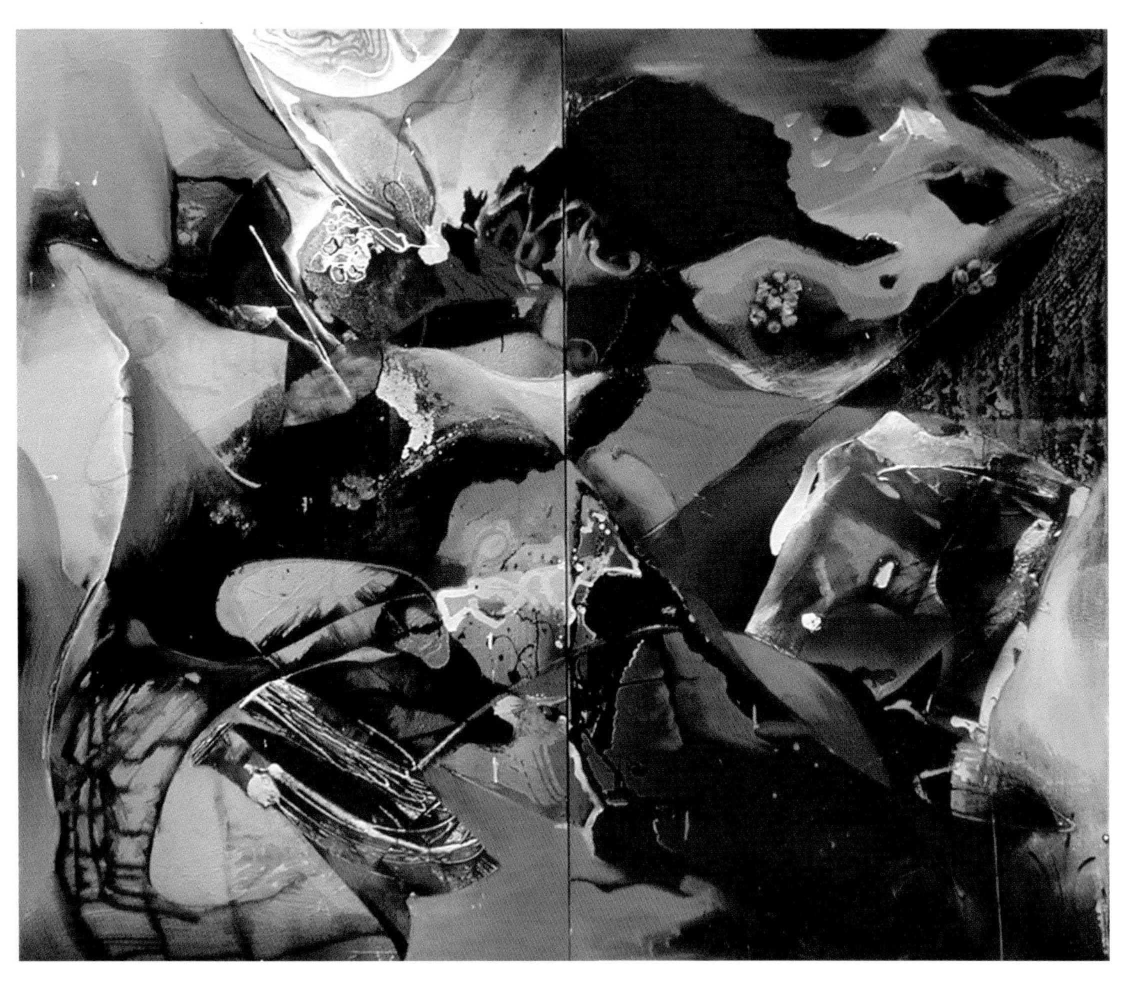

South Way, Cristina Popovici, 2002.
Courtesy Cristina Popovici

Romania and spent six years in Christchurch before moving to Auckland. She paints in a completely non-representational style, producing vibrant, colourful images. *South Way* was inspired by this artist's feelings about the Canterbury high country, with its range of colours, shapes and moods.

Big-screen high country

Millions of cinema-goers around the world have been fascinated by the awesome scenery used by Peter Jackson as the backdrops for his *Lord of the Rings* film trilogy. One eye-catching setting was the limestone rocks of Flock Hill in the Castle Hill Basin; in the upper Rangitata Valley the village of Edora was built on top of Mt Sunday, and the valley across the river was the background for Helms Deep. Several travel companies now offer sightseeing trips to these locations. Flock Hill was also used for the battle scenes in Andrew Adamson's *The Chronicles of Narnia: The lion, the witch and the wardrobe.*

Canterbury high-country localities also appear in the background of television commercials and print advertisements, where they are often used to suggest a rugged, pristine environment.

Across the Rangitata River from Erewhon is this narrow valley, used to spectacular effect in the *Lord of the Rings* film trilogy as the setting for the fortress of Helms Deep.

Courtesy 4 x 4 New Zealand

BIBLIOGRAPHY

Acland, L. G. D. *The Early Canterbury Runs*, Whitcombe & Tombs, 1946.

Ainsley, Bruce. 'Heart in the Highlands', *NZ Listener*, 12 July 2003.

Ainsley, Bruce. 'This Land is Your Land', *NZ Listener*, 2 March 2002.

Anderson, Mona. *The Good Logs of Algidus,* A. H. & A. W. Reed, 1965.

Anderson, Mona. *A River Rules My Life*, Whitcombe & Tombs, 1963.

Barker, Lady. *Station Life in New Zealand*, Whitcombe & Tombs, 1950.

Barker, Lady. *Station Amusements in New Zealand*, Whitcombe & Tombs, 1950.

Barnett, Shaun. 'Across the Alps', *NZ Geographic*, No. 47, July 2000.

Berendstrum, Eric. 'Canterbury's Damaging Nor-Wester', *NZ Geographic*, No. 1, January 1989.

Bestic, Kim. 'Forgotten Plants of the Shingle Riverbeds', *Forest & Bird*, No. 311, February 2004.

Bishop, Nic. *Natural History of New Zealand*, Hodder & Stoughton, 1992.

Booth, Robert. *Five Years in New Zealand (1859 to 1864)*, J. G. Hammond, 1912.

Brabyn, Sven, and Elsie Bryant. *Tramping in the Southern Alps: Arthur's Pass to Mt Cook*, 1994.

Brailsford, Barry. *Greenstone Trails*, A. H. & A. W. Reed, 1984.

Britten, Rosemary. *Lake Coleridge: The power, the people, the land*, Hazard, 2000.

Burrows, Colin J. *Julius Haast in the Southern Alps*, Canterbury University Press, 2005.

Butler, Samuel. *A First Year in the Canterbury Settlement*, Blackwood & Janet Paul, 1964.

Cant, Garth, and Russell Kirkpatrick (eds). *Rural Canterbury: Celebrating its history*, Daphne Brassell Associates and Lincoln University Press, 2001.

Chapman, John. *The Stations of the Ashburton Gorge*, 1999.

Churchman, Geoffrey B. *The Midland Line: New Zealand's trans-alpine railway*, IPL Books. 1988.

Churchman, Geoffrey B., and Tony Hurst. *The Railways of New Zealand: A journey through history*, Collins, 1990

Coates, Glenn, and Geoffrey Cox. *The Rise and Fall of the Southern Alps*, Canterbury University Press, 2002.

Cobb, Len, and James Duncan. *New Zealand's National Parks*, Paul Hamlyn, 1980.

Cooke, John. 'When the Kb Class Ruled the Midland Line', *NZ Railfan*, June and September 2000,

Cooke, John. 'Kb Class Locomotive Scrapbook', *NZ Railfan*, March 2001.

Cooke, John. 'Allie Godbaz, Engine Driver', *NZ Railfan*, March 2002.

Craig, Martin. 'High Country at Low Prices', *Fish & Game New Zealand*, No. 17, 2003.

Crawford, Sheila S. *Sheep and Sheepmen of Canterbury, 1815–1914*, Simpson & Williams, 1949.

Darby, John et al. (eds). *Natural History of Southern New Zealand*, University of Otago Press, 2003.

Druett, Joan. *Exotic Intruders*, Heinemann, 1983.

Ell, Gordon. 'Protecting the High Country', *Forest & Bird*, No. 305, August 2002.

Ensor, Peter C. *Many Good Years, Some Not So Good: A history of the Double Hill Station*, P. C. Ensor, 1990.

Gage, Maxwell. 'Late Pleistocene Glaciations of Waimakariri Valley', *NZ Journal of Geology and Geophysics*, Vol. 1, 1958.

Gage, Maxwell. *Legends in the Rocks: An outline of New Zealand geology*, Whitcoulls, 1980.

Gage, Maxwell. 'On the Origin of Some Lakes in Canterbury', *NZ Geographer*, Vol. 15, No 1., 1959.

Graeme, Ann. 'Hidden Worlds of Ashburton Lakes', *Forest & Bird*, No. 305. August 2002.

Grzelewski, Derek et al. 'Salmon: The miracle fish', *NZ Geographic*, No. 63, May–June 2003.

Handbook to the Arthur's Pass National Park, Arthur's Pass National Park Board, 1986.

Holm, Janet. *Nothing But Grass and Wind: The Rutherfords of Canterbury*, Hazard, 1992.

Homer, Lloyd, and Les Molloy. *The Fold of the Land: New Zealand's national parks from the air*, Allen & Unwin/DSIR, 1988.

Kennaway, L. J. *Crusts: A settlers fare due south*, Capper Press (reprint), 1970.

Logan, Robert. *Waimakariri: Canterbury's river of cold rushing water*. R. Logan, 1987.

McKinnon, Malcolm (ed.). *Bateman New Zealand*

Historical Atlas, Bateman/Department of Internal Affairs, 1997.
McLeod, David. *Down from the Tussock Ranges*, Whitcoulls, 1980.
McLeod, David. *New Zealand High Country*, Canterbury Agricultural College, 1951.
McNeish, James, and Gareth Eyres. 'The Man Between the Rivers', *NZ Geographic*, No. 8, October–December 1990.
Making of NZ: Pictorial surveys of a century, No. 1, The Beginning; *No. 7, The Squatters*; *No. 16, Tracks and Roads*, Department of Internal Affairs, 1939.
Mankelow, Sarah. 'Treasures of Castle Hill', *Forest & Bird*, No. 299, February 2001.
Mirfin, Zane et al. *Brown Trout Heaven*, Shoal Bay, 2000.
Mitchell, Dave. 'Biking the Craigieburn', *Wilderness*, September 2004.
Morris, Jim. *Different Worlds: Backcountry yarns.*
Nell, Lyn, and Judith Stratford (eds). *West of Windwhistle: Stories of the Lake Coleridge area*, Lake Coleridge Tourism Group, 2005.
Newell, Claire, and Leith Newell. *Castles in the Air: The early history of skiing at Broken River*, 2001.
Newton, Peter. *Mesopotamia Station: A survey of the first hundred years*, Timaru Herald, 1960.
Newton, Peter. *The Boss's Story: The problems and pleasures of managing a New Zealand sheep station*, Reed, 1966.
Power, Gerry. *White Gold: The Mount Hutt story*, Canterbury University Press, 2002.
Relph, D. H. 'The Canterbury Nor-wester', *NZ Geographic*, No. 70, November–December 2004.
Relph, D. H. 'A Century of Human Influence on High Country Vegetation', *NZ Geographer*, Vol. 14, No. 2, October 1958.
Relph, D. H. 'The Vegetation of Castle Hill Basin', *NZ Geographer*, Vol. 13, No. 1, 1957.
Richards, E. C. *Castle Hill*, Simpson & Williams, 1951.
Robertson, Hugh, and Barrie Heather. *The Hand Guide to the Birds of New Zealand*, Penguin, 1999.
Ryan, A. P. *The Climate and Weather of Canterbury*, NZ Meteorological Service, 1987.
Schaer, Cathryn. 'Little Pistes of Heaven', *Air New Zealand Magazine*, July 2005.
Scott, Mark, and Peter Quinn. 'Arthur's Pass: Heart of the mountains', *NZ Geographic*, No. 52, July–August 2001.
Seitzer, Stefan, and Gordon Roberts. 'Thar and Chamois: Monarchs or misfits?', *NZ Geographic*, No. 23, July 1994.
Stupples, Polly. 'Fields of Gold', *NZ Geographic*, No. 66, November 2003.
Soons, J. M. and M. J. Selby (eds). *Landforms of New Zealand*, Longman Paul, 1990.
Soons, J. M. 'Survey of periglacial features in NZ: Land and Livelihood' in *Geographical Essays in Honour of George Jobberns*, NZ Geographical Society, 1962.
Soons, J. M. 'The Geomorphology of the Cass District', in C. J. Burrows, *Cass*, Department of Botany, University of Canterbury, 1977.
Spirit of the High Country: The search for wise land use, South Island High Country Committee of Federated Farmers, 1992.
Stevens, Graeme. *Prehistoric New Zealand*, Heinemann Reed, 1988.
Thornton, Jocelyn. *Field Guide to New Zealand Geology*, Heinemann Reed, 1985.
Todhunter, Ben. 'The Future for High Country Farmers: The station owner's view', *Primary Industry Management*, Vol. 7, No. 3, September 2004.
Tussock Grasslands: Our heritage, South Island High Country Committee of Federated Farmers, 2001.
Walrond, C. 'The Rocks of Castle Hill', *NZ Geographic*, No. 44, October–December 1999.
Wodzicki, K. *Introduced Mammals of New Zealand*, DSIR, 1950.
Woods, John. 'Masochism & Magic', *NZ Geographic*, No. 25, January–March 1995.

Useful websites

The Internet has become a major source of information about the Canterbury high country, particularly for recent developments such as tourism and issues of land management and conservation. The following is a list of websites that have provided useful information about this region and were current in 2006.

Tourist accomodation

Waimakariri

Arthur's Pass: Mountain House: www.trampers.co.nz
Bealey Hotel: www.bealeyhotel.co.nz
Castle Hill: www.theburn.co.nz
Cora Lynn: www.wildernesslodge.co.nz
Grasmere: www.grasmere.co.nz
Flock Hill: www.flockhill.co.nz

Rakaia
Ryton: www.ryton.co.nz
Terrace Downs: www.terracedowns.co.nz
Lake Coleridge Lodge: www.lakecoleridgelodge.co.nz

Rangitata
Mesopotamia: www.mesopotamia.co.nz
Mount Potts: www.mtpotts.co.nz
Stronechrubie: www.stronechrubie.co.nz

Safari hunting, fishing, ecotourism and other guided activities
Back Country New Zealand (Methven): www.backcountry.co.nz
4 x 4 New Zealand (Geraldine): www.4x4newzealand.co.nz
Tussock & Beech Ecotours (Staveley): www.nature.net.nz
Snow & Stream Lodge (Methven): www.fishandhunt.co.nz
Lake Coleridge Tourism Group: www.lakecoleridgenz.info
Craigieburn Alpine Safaris: www.craigieburn.com
Hunt NZ Safaris: www.huntnewzealand.com
Four Seasons Safaris: www.hunting-fishing.co.nz

Jetboating
Waimakariri: www.waimakalpinejet.co.nz
Rakaia: www.rivertours.co.nz

Skiing
Broken River Ski Club: www.brokenriver.co.nz
Craigeburn Valley Ski Club: www.craigieburn.co.nz
Mt Cheeseman Ski Club: www.mtcheeseman.com
Mt Hutt ski area: www.nzski.com
Mt Olympus ski area: www.mtolympus.co.nz
Mt Potts Backcountry: www.mtpotts.co.nz
Porter Heights ski area.www.porterheights.co.nz
Mt Hutt Helicopters: www.mthuttheli.co.nz

Adventure activities
Speights Coast to Coast multisport event: www.coasttocoast.co.nz
Back of Beyond mountain bike tours (Ashburton): www.mountainbiking.net.nz
Aoraki Balloon Safaris (Methven): www.nzballooning.com
Arthur's Pass walk and camping: www.onenightoutdoors.com
Arthur's Pass mountaineering conditions: www.softrock.co.nz

Land management and conservation
Department of Conservation: www.doc.govt.nz
Arthur's Pass National Park region: www.apinfo.co.nz
Royal Forest and Bird Protection Society: www.forestandbird.org.nz
Land Information NZ: www.linz.govt.nz
Natural history in Arthur's Pass National Park: www.natureandco.co.nz

Other useful sites
Institute of Professional Engineers (information about the Midland Railway line): www.ipenz.org.nz
Weather and climate information: www.metservice.com; www.niwa.co.nz

INDEX